Thriving
During
Challenging Times

Thriving
During
Challenging Times

The Energy, Food and Financial
Independence Handbook

Cam Mather

AZTEXT
PRESS

Aztext Press
Tamworth, Ontario Canada K0K 3G0
michelle@aztext.com • www.aztext.com

Library and Archives Canada Cataloguing in Publication

Mather, Cam, 1959-
  Thriving during challenging times : the energy, food and financial independence handbook / Cam Mather.

Includes bibliographical references.
ISBN 978-0-9733233-6-8

  1. Self-reliant living. 2. Home economics. 3. Finance, Personal. 4. Cost and standard of living. I. Title.

GF78.M38 2009 640 C2009-902811-5

Printed and bound in Canada

This book is dedicated to my wife and soulmate Michelle who for 33 years has tolerated the ups and downs of a restless soul, who finally found what he was looking for, off the grid in the woods north of Kingston.

Acknowledgements

This book was very much a collaboration with my wife Michelle who helped research much of the material, offered wisdom and guidance during my constant diatribes about how the book was coming together, and spent endless hours editing the various drafts which set a new standard for how long sentences can run on or how short sentence fragments can be. Someday I hope my fingers on the keyboard will be properly wired to my brain to make Michelle's job a little easier.

I would like to thank my uncle Ian Micklethwaite who was a mentor to me early in my business career and helped me to find a way to turn my passion for living sustainably into a way to earn my living.

I am indebted to my daughters Nicole and Katie who tolerated my restless search for meaning and independence in life which in their formative years took them many miles from the nearest mall. Katie also provided exceptional assistance in the final stages of the book, especially in reference checking.

Ten years after arriving in the middle of nowhere we are still here thanks to the greatest neighbors and friends on the planet, Ken and Alyce Gorter. The key for a cidiot (city idiot) like me to move off the grid successfully is to have a neighbor like Ken who is a maestro of all things practical from electricity to concrete to steel. The key for a gardener attempting to grow in sandy soil is to have a neighbor like Alyce who is happy to keep the horse manure coming year after year and to school us in the ways of the country.

Our move to the middle of nowhere was helped by the loyalty and support of our good friends in Burlington Karen and Pete Dougherty and John Wordsworth.

Much of the continuous quality improvement of our renewable energy system has been facilitated by Jerry Horak, who can disassemble an inverter room and put it back together better than anyone else I know. Ellen Horak provided the most detailed critical evaluation of the preliminary draft of the book and offered exceptional guidance to get it to its final form.

Deborah Sowery-Quinn provided great insight into the book when she read an early draft, and I greatly appreciate her feedback.

I'd like to thank my cousin David Flett for his help with some of the financial sections and for his constant music trivia emails that help keep me sharp when it comes to the ever-important world of contemporary music lyric trivia.

Our editor Joan McKibbin not only provides exceptional editing, she is the coolest music-loving, chainsaw-wielding, computer-instructing, animal-loving editor on the planet. Now if we can only get her to change her light bulbs to compact fluorescents.

Finally our pursuit of making a living promoting a sustainable lifestyle would not have been possible without the patience, guidance, and exceptional work of author and clean-energy advocate Bill Kemp and his wife Lorraine. I hope someday to ultimately grasp the concept of the difference between power and energy. In the meantime I will continue to arrive at our meetings with my renewable energy "how to...?" questions until I get it.

Table of Contents

PART I

Chapter 1	Introduction	1
Chapter 2	How We Got Independent	5

PART II   Challenges

Chapter 3	Economic Collapse	9
Chapter 4	Peak Oil	23
Chapter 5	Peak Food	33
Chapter 6	Peak Water	41
Chapter 7	Climate Change	47
Chapter 8	Why It's Different This Time	51

PART III   Solutions   59

Chapter 9	Where to Live	63
Chapter 10	Where to Work	83
Chapter 11	Heating	95
Chapter 12	Fuel/Energy in Your Home	107
Chapter 13	Food	133
Chapter 14	Gardening	155
Chapter 15	Water	183
Chapter 16	Transportation	193
Chapter 17	Health Care	207
Chapter 18	Safety and Security	223
Chapter 19	Money	233
Chapter 20	Mediums of Exchange	263
Chapter 21	Living Happily in the New World	283
Chapter 22	Thriving During Challenging Times	299

PART I

Introduction

"Toto, I don't think we're in Kansas anymore."
Dorothy, from The Wizard of Oz

The world just changed.

Well of course it changed. It's constantly changing. It's complex. Six and half billion people live here. Of course it's changing.

No, I don't mean in the usual ways.

I mean the reality that most North Americans face.

I mean the world in which we find ourselves today is not one most of us have ever experienced. Many of us won't like this new world. We'll long for it to return to the way it was. The cheap energy. The abundant products to buy. The infinite food choices. The luxury of cheap gas and independent personal transportation that allowed us to go anywhere we wanted, whenever we wanted. And the stable economy and jobs that went with it, that gave most of us a high enough income that we could enjoy all these luxuries.

"It'll go back to being the way it was before, right?" It would be nice. We all wish it would, but this time is different. This time our version of reality has been altered, and this time it's likely to be permanent. We may go through some good times, or better times, where things seem like they are getting back on track, but there's a good chance these will be short-lived.

I wish things would go back to the way they were, but I don't think they are going to. I think this time, it's different.

For a decade I have been giving workshops on living with renewable energy and over the last three or four years I have broadened them to deal with the many challenges we're facing. People are often shell-shocked because it is a lot to absorb in an afternoon. This is where this book came from. It will help me provide more detail to a larger audience. At the end of the book I provide a bibliography of books I've read over the last five years that has shaped my world view. I would strongly recommend you use this book as a starting point for further reading on the many challenges the world faces.

As I've become more aware of these multiple challenges I've also been inspired to make myself more independent and I've found a huge benefit: it makes you feel really good. It gets priorities back in balance and lets you focus on what's important. For some reason I was inspired to plant a vegetable garden in my parents' suburban home when I was 16, and I see now 34 years later that it was the start of a journey to find meaning that you simply can't find watching television or surfing the net. I have never earned less money in my adult years than I do right now, and I've never been happier and more at peace. I hope to be able to provide you with the tools to attain that same feeling of contentment.

So first I'm going to explain what happened. How did it happen that things went from being so good to being, well, not-so-good today. Once you understand how it happened, hopefully it will be a little easier to deal with the new reality. To many people, this is not a surprise. Many people were saying for many years that things weren't right and that the problems we are experiencing right now were going to be the result of not altering our behavior.

Most of us know this from our weather payback theory. "It's been a great fall. Oh, we're going to pay for it this winter."

Then once we've examined how this happened, I'll try and convince you why it's unlikely things will go back to the way they were before. With 6 1/2 billion people on the planet, we have hit overshoot, where we've gone past the ability of the world to support so many people. The days of cheap food and cheap energy are behind us. The days of abundance are gone. We've now entered the time of scarcity.

This does not have to be a bad thing though. I hope to show you how to deal with the new economic reality in ways that can actually make you feel better. I'll show you how, by developing a strategy to deal with the new world, many of the new things you'll be doing will help you not only live in this new world of scarcity but also reduce your footprint on the planet and make you feel better. It will be better for your soul.

At the heart of this book therefore is a message of hope. Things are different than they were, but they're not necessarily worse. It's how you look at them that matters. It's how you deal with the new world. You need to rethink everything you do in terms of economic gain and start thinking in terms of how it will help you live today.

We were conditioned for many decades to be obsessed with saving for tomorrow, putting large sums of money aside for our retirement. Many people who did retire were miserable. They have lots of money but not enough to do and not enough to give their life purpose. The modern economy and democratic society have basically removed any of the need for humans to have to worry about the most basic of human needs, like eating, or staying warm, or surviving for one more day. On one level this was a very good thing. It was a huge accomplishment in terms of human evolution, but the systems we created allowed the species to get lazy and many of us ended up miserable. It took away many of the experiences that have given human lives meaning and joy.

So now it's time to deal with the new reality. For many it won't be easy. Many of us feel overwhelmed. We feel dazed and confused, as if someone has just punched us in the face. The sort of economic dislocation that is going on right now is nothing short of a full knock-down, drag-out body blow. Feeling disillusioned is normal. Feeling sad is normal. Feeling discouraged is natural. But at a certain point, it's not healthy. Eventually you've got to come to grips with the task at hand and get moving.

The movie *The Shawshank Redemption*, which was adapted from a Stephen King short story, takes place in a prison in an environment of terrible oppression. At a critical point the lead character played by Tim Robbins is at his wit's end, having lost hope. The character played by Morgan Freeman tells him he has a choice to either "Get busy livin' or get busy dyin'." It's a choice. It's basically a choice you've got to make now. "Do I let this big change take me down with it, or do I rise to the challenge and start dealing with it?" Do I give in to the laziness of despair, or do I use this as a challenge and rise to the occasion and see it as a huge opportunity for new meaning in my life?

Living in the new world is not going to be worse than it was, it's just going to be different. Humans are adaptable creatures, so now it's time for you to adapt.

I think the new time will be a time of great potential, where humans learn different ways of dealing with challenges. You have within you the potential to rise to this challenge and thrive. You can create a new reality for yourself, one where you'll be able to take more joy in the simple things. The simple act of growing your own food and preparing it. The simple act of harvesting your rainwater to use in your home. These are good things. These can bring great satisfaction and can fill your soul with joy.

There are many books on the various challenges we face. They are often big picture views of the problem and do not offer a practical strategy for an individual to use. They end up being kind of discouraging because they love to point out the problems, but one is not always left with a feeling of empowerment as to how to deal with them. One line of reasoning with many of these books is that there will be such a complete collapse that you'll have to live in the woods and scavenge as humans did in the early stages of our evolution. The other suggests loading up on canned goods and guns and having the SUV fully gassed up so you can bug out when things get bad. Bug out to where? If things are that bad there's not going to be gas to fill that gas tank up with. I don't believe things will get that bad, and I don't agree with either of these lines of thinking.

This is not a big picture book. It's a little picture book. It's a "you" book, showing "you" how to deal with the problems. The problems are big and sometimes seem insurmountable, but I believe if we all make the move to independence—in our food production, in our energy use, and in living within our means financially—many of the larger environmental and economic problems will look after themselves. It's this personal self-determination that offers the solution.

This is the reason I've structured the book by outlining the challenges in the first part of the book and offering solutions later. There are lots of challenges, and if I had put the challenge in each chapter I think it might have been a bit discouraging. I think it's a better approach to just lay out what's happening and then give you lots of solutions so you can get on with dealing with them. It's like taking a bandage off quickly rather than prolonging the discomfort. This book is based on workshops I've been giving for years and I've found it's best to follow this approach. By the end of the day people are motivated, have the tools, and are ready to get on with implementing their strategy to deal with the challenges.

So let's get on with it. First we'll find out how we got into this mess; then I'll give you the tools you need to come up with a strategy that is going to help you deal with it, rise above it, and be happy—really happy while you're at it!

2 How We Got independent

My wife Michelle and I live on 150 acres in the woods in Eastern Ontario in Canada. During our first seven years here our daughters, who are now 21 and 23, lived with us. They've both since moved out and are now living in the city. We are surrounded by thousands of acres of bush that are partly owned by groups of men who use it for two weeks of the year for deer hunting, and the rest is an undeveloped provincial park. Our nearest neighbors are 2 ½ miles (4 km) to the east and 4 miles (6 km) to the west. This happens to be where the electricity poles end as well, so we are completely off the electricity grid and generate all our own electricity with solar and wind and an ever-decreasing amount of fossil fuel in the form of gasoline that we burn in our generator.

We moved here from a busy suburban street three hours away in Burlington, a suburb near Toronto, Ontario. Our journey to this place in the bush actually began over 25 years ago when we first considered the idea of recycling. We lived in an apartment in a medium-sized city, and I was looking at our trash one day and thought, a lot of energy has gone into making these cans and newspapers and glass that we're throwing out. Surely there has to be a way to do something with them.

We discovered a small group of like-minded people and began accumulating our recyclables and driving them to an industrial warehouse so they could be sold into the fledgling commercial recycling market. This was long before curbside pickup of recyclables.

A few years later we were living in our own home and the city had begun taking away our recyclables from the curbside. By then we had turned our attention to working to reduce our trash, getting it down to one can of garbage every eight weeks for our family of four. We were active in the local environmental group "The Conserver Society" and I served on the City of Burlington's "Sustainable Development Committee." Still, something seemed to be missing. We had a small house with no air conditioning and large black walnut trees that helped keep the house cool. Nothing grew under those trees in our backyard, so we had vegetables growing in the front yard.

I think it was after a visit to my chiropractor who asked, "Cam, was that corn I saw growing in your front yard?" that we decided it was time to get out of the city. The pace was too fast, there was too much of a focus on buying stuff and accumulating stuff, and as motorists got more stressed and switched to larger, wider, more dangerous SUVs, cycling was starting to feel just too risky.

So we established a five-year plan to move to the country. We put together a wish list for our dream home and started looking. After much time and many wasted trips we finally found it. We ran an electronic publishing business at the time and it seemed as though most of our customers would stick with us if we moved. Technology had allowed us to provide most of our services electronically, which made face-to-face meetings less necessary.

Moving three hours away from our customers was a scary proposition. Moving to an off-grid house with absolutely no clue about electricity was terrifying. Yet like many of life's defining moments, the greatest challenges presented the greatest opportunities. There were many hiccups along the way. It hasn't been smooth sailing, and there have been moments when Michelle and I have looked at each other and said. "Well, do we throw in the towel and move back to the city?" But they have been few and far between, and after we've taken a moment to re-evaluate the challenge of the moment we've figured out a strategy to deal with it.

Our learning curve has been huge. We went from not really knowing the difference between AC and DC electricity to doing workshops across the province that allow us to share what we've learned and inspire others to integrate renewable energy and sustainability strategies into their own lives.

We've learned about harvesting our own firewood, dealing with water issues in a rural home, and how much work is involved with growing much of your own food. There have been many successes and as many failures.

The one overlying theme in our journey to independence has been what we tried to instill in our daughters while we home schooled them, and that's the concept of lifelong learning. Education does not end when you leave public school or complete your university or college degree—in fact that's just the beginning. Each of our challenges, whether it was adding more solar panels, putting up a wind turbine, or installing a solar thermal system to heat our hot water, has seemed complicated and often well beyond our skills at the beginning. But we read all the information we could find, consulted with everyone we could, and finally took the plunge and did it. We have had many great teachers along the way who've taken the time to share their knowledge and we are extremely grateful to them.

Now we want to share what we've learned. It comes from a lifetime of learning about how to be more independent. It comes from a lifetime of research into the challenges that we face as a species and the steps you can take to prepare yourself for these changes. It also comes from a realization that this move to independence and accomplishing these tasks on your own creates an incredible sense of well-being and happiness. Years after putting up my wind turbine I still stand under it and watch it endlessly producing clean green renewable energy for my home and it feels as if my chest will explode with pride. It's way better than TV!

What you need to do is to start creating an independent living space now to ride out any storms that may be coming. The technology exists to keep you living a pretty comfortable life even if the electrical utility is having trouble keeping the power on. At our house there are no power lines or utility poles of any type bringing energy to our home. We are almost completely independent in terms of our energy, food, and income. If need be, we could be completely independent and our lifestyle would not notice a real decline. This is the goal you should be striving for, and the technology and knowledge are here, today. They're in this book!

There's never been a better time to get started on your own journey of discovery about the challenges we face and the steps you can take to prepare. It can be a wild ride, but it's a blast!

PART II
Challenges

3   Economic Collapse

Millions of Americans are behind on their mortgage payments and millions more have lost their homes. Pension funds have been decimated, retirement plans put on hold, the economy is shedding jobs in the hundreds of thousands every month. Our industry sector is in free fall; General Motors, formerly the largest, most profitable corporation on the planet, is bankrupt; most states face crippling deficits; and California, the seventh-largest economy in the world, is teetering on insolvency.

So how did this all happen? How did we end up in this mess? Things weren't supposed to be like this. Things were supposed to continue as they always were, a continuing higher standard of living, bigger houses, and a better life for everyone.

Was it the sub-prime meltdown? The housing bubble? Collateralized Debt Obligations (CDOs)? Asset-Backed Commercial Paper? hedge funds? derivatives? credit default swaps? Never before have there been so many new financial instruments that seemed to suddenly be responsible for sending our economy into a tailspin.

These may have been the straw that broke the camel's back, but the roots of the current economic challenges go way back.

A short history of money is a good first step.

The Gold Standard

Once humans got organized enough to start working together, they learned it was advantageous to trade. If one person or group fished well and another farmed well, why not trade the excess and vary your own diet. This worked in many facets of human endeavor.

As humans began trading further afield, it often became inconvenient to barter directly with another group or individual. Sometimes it was better to let individuals who traveled take your goods with them. However, you needed a way for them to compensate you for the goods when you relinquished them rather than after they came back, because who knows if you were ever going to see them again.

Precious metals were the solution. Gold and silver had begun to hold value in terms of jewelry and ceremony, so they had intrinsic value. Now the trader could purchase your goods from you with gold and travel elsewhere with them while you received payment immediately.

Archeologists have found gold and silver coins that are thousands of years old, so humans discovered this amazing medium of exchange a long time ago. The Roman Empire was known for its gold coins and the value they held throughout the empire.

Eventually, as the world developed and economies grew, the volume and value of goods expanded and carrying and transporting large quantities of gold coins became cumbersome. Humans needed another way to trade.

What was created was what we know today as dollar bills. Originally these would have been more of a check or note, but essentially it was a piece of paper that said the bearer of the note could redeem it for a specified amount of gold. Like a check, it was a claim against someone's bank account or precious metals deposits. This way you didn't have to lug around pounds of gold.

In England the "pound" is used rather than the dollar, and "pound" actually refers to pound sterling, since early in its history a pound note could be redeemed for a pound of silver. This paper substitute worked well on a number of levels. As long as you had confidence that the person who issued the note was good for the money, it was far more convenient.

Eventually though, as humans always do, someone found a way to corrupt the system. Many times it was governments. As soon as they found that they could in essence create money out of thin air by printing a piece of paper, they often couldn't resist. Many European governments were notorious for living beyond their means by doing this until someone finally called them on it and said, "OK, I want to redeem the gold that these paper notes are supposed to represent." When they discovered the castle vault didn't have the gold to back it up, systems broke down.

As society became more organized and groups started forming and calling themselves countries with governments, currency and trading became more complicated, but it still all boiled down to gold. The government needed gold sitting in a vault somewhere before it could issue paper notes or currency. This was what was known as the "gold standard." Over time it drifted in and out of fashion, but ultimately it was one of the most effective systems of economic exchange.

Sometimes governments issued what was called a "fiat currency" which was not backed by gold. Its value was based on the government of the time saying it had value. "Trust me, this piece of paper has value, because I say it does." Sometimes this worked. Sometimes it didn't. If the government was a stable one that collected taxes and was responsible, sometimes a fiat currency worked. But the key to it working was confidence. As long as everyone had confidence that a piece of paper had a value that could be counted on, they were willing to exchange their labor or their goods for it. If the government did something to undermine that trust, the fiat currency quickly became worthless.

This was where the gold standard worked better. It removed the variable of human weakness. It ensured that the currency had value because it was backed by gold. Even if the king was a crackpot, as long as he left the gold in the treasury the paper currency was fine.

When we entered the 20th century, developed countries went through various iterations of the gold standard. After World War I, Britain had to ship huge amounts of gold to the United States to compensate them for their help in defeating the Germans. After World War II, with many economies in shambles, the Bretton Woods Agreement was meant to get the world back on track by having governments stick to the gold standard. At that time the U.S. was flush with gold, stored in Fort Knox. Obviously the U.S. wanted the world to follow the gold standard, because it was working very well for them.

One of the nice things about the gold standard was that it kept governments in check. It was similar to your personal line of credit or ability to take on debt. In our households we have an income and we have assets. Sometimes we wish to purchase things that are beyond what we have in our bank accounts. When evaluating a mortgage application the bank looks at the value of the house and how much the applicant is earning and decides whether or not the applicant is a good credit risk and will be able to pay that mortgage amount back. We can also get credit cards that allow us to purchase goods when we don't have the cash in our wallets. With the credit card we're given a credit limit. When we exceed that credit limit, the credit card company cuts us off and says we can't use it for transactions anymore.

At a macroeconomic level, the gold standard served as a credit limit for countries. When they traded with each other, they settled up with gold. As long as you were trading with a country that you knew was good for the gold, things went well. But if a country was living beyond its means, spending more than it was earning, and was not able to back up its purchases with something of value—either a solid economy or a central bank with lots of gold reserves—other countries would cut them off. It would be like the U.S. government getting to the counter to pay for that new plasma TV and the credit card company declining the purchase. And we all know how embarrassing that is.

So after World War II the United States, which was the largest economy in the world, used the gold standard. And it worked pretty well. We had a few decades of pretty positive times. But the U.S. got bogged down in the quagmire of Vietnam, and by the time Nixon was President the country's finances were a mess. So Nixon had two choices. The first, which he didn't relish, was to declare bankruptcy. That's what it amounted to. He would have had to tell the world that the U.S. didn't have the gold to back up the debt it had run up and was defaulting on its loans. Obviously this was not a very palatable option. It certainly would have been the correct thing to do— to deal with the problem then and not postpone it—but he took the more politically expedient route.

In 1971 Richard Nixon declared to the world that the United States was going off the gold standard. From then on, he told the world, the U.S. dollar would be the standard of exchange worldwide. It would become the "reserve currency" of the world. Gold was no longer relevant and the U.S. fiat currency, the "greenback," was the standard of exchange. And it had value, because the United States said it did. And granted, at the time there was a lot behind the U.S. dollar. The GDP of the country was monstrous compared to the rest of the world. The U.S. had significant oil reserves and a very large, well-developed military. To the world, it looked as though the U.S. dollar was a pretty good alternative to the gold standard.

The U.S. went one step further and told the world that anyone wanting to trade oil needed to do it in U.S. dollars. The U.S. was an energy powerhouse, producing and purchasing huge amounts of oil, so this seemed logical to the world as well. So if you were a country without oil and you needed to purchase some, you needed to get some U.S. dollars. You needed to trade something with the U.S. to get those dollars. Obviously this system worked very much to the benefit of the U.S.

But like all fiat currencies, whether they are with drunken debaucherous kings or democratically elected central governments, as soon as a government discovers that all it has to do to create more money is to crank up a printing press, no government has ever declined to do so. It just makes sense. If it were you, you'd do the same thing. How many people say that if they had a genie's magic lamp and the genie granted them one wish, they'd wish for a million more wishes. It's something for nothing.

And this is where the current economic mess all started. The U.S. was living beyond its means. As the U.S. government debt grew, economists said it was sustainable because as long as the GDP continued to grow, the debt could be whittled away.

When Nixon left office the U.S. debt was about $500 billion, or half a trillion dollars. The GDP was about $1.5 trillion, so the debt was about 35% of the GDP. As of 2009 the U.S. debt is $12 trillion dollars. The annual U.S. economic output is around $14 trillion, so the current national debt-to-GDP ratio is closer to almost 90%. It is a staggering, mind-boggling number. On March 16, 2006 the U.S. Congress voted to increase the national debt by $781 billion dollars because previous Congresses had set a limit on how high the U.S. debt could go. Rather than reining in spending or increasing taxes, it took the easy route and just raised the limit. With a population of 305 million people, each U.S. citizen's share of the debt is almost $37,000.

When George W. Bush took office in 2000, he inherited a budget surplus, which meant that the federal government took in more than it spent. This was after eight years of "tax and spend" democrats under Bill Clinton. George Bush decided to take that surplus and, rather than applying it against the $5.6 trillion debt, introduce tax breaks, many of which were of great benefit to higher-income Americans. In the meantime, after September 11, 2001, he cranked government spending. So he not only reduced taxes, he increased spending. If you reduce taxes, you need to reduce spending. Unless of course you never had any intention of paying back your debt. And with the U.S. debt ballooning to almost $10 trillion dollars under his watch, it looked very much as though he had no intention of paying it back.

People often go bankrupt when they run up too much debt. With mortgages and credit card debt, sometimes people spend more than their income allows them to, and they have to declare bankruptcy. In the U.S., often a serious medical issue causes this when the person does not have proper health care insurance coverage. It's a painful and humiliating experience for many who have the best intentions, but circumstances simply catch up to them. Others, though, are quite smart about it. When they realize bankruptcy looms, they ratchet up spending and give much of the loot away to friends and family, knowing they're never going to have to repay it. This seemed to be the intention of the second Bush presidency. If he did think anyone was ever going to pay back the debt, he must have known it was going to require unbelievable pain and disruption to the American people. But he was happy as long as it happened on someone else's watch.

Most people try and live within their means. We all enjoy living in a house we own, even if we don't have the cash to pay for it when we're young. We take out a mortgage that we pay back over time—20 or 25 or 30 years. But we have every intention of paying it back. For some reason the government of the United States apparently decided that it would never bother repaying its debts. Politicians don't even want to talk about it. When then Controller General David Walker was interviewed by CBS News on *60 Minutes* about the mess of U.S. Government finances, no elected Senator or Member of the House of Representatives would appear on camera. These are the people elected to run the government on the people's behalf, but they wouldn't even talk about the financial mess.

So how did the U.S. get so far in debt? If it didn't have gold to back up the money it issued, who was giving the country credit? Well, apparently most of the world. Central banks didn't mind having U.S. dollars in their vaults instead of gold. Other countries wanted to trade with the U.S., so when they shipped their goods to America, they got paid in U.S. dollars. For most countries it was not advantageous to convert the dollars into their own currency, so they had to do something with it. During the 1990s, many of those dollars ended up back in the U.S. stock market. The Dow during the 1990s went through unprecedented growth, opening the decade at 2,750 and ending at 11,500. Books were written projecting it would hit 39,000. Much of this growth was precipitated by U.S. dollars flowing back into the stock market.

The NASDAQ stock exchange, which represented many of the new high tech and emerging companies, also experienced unprecedented growth. Eventually it became known as the "Tech Bubble" when the value that was assigned to companies with products that had yet to prove themselves as being able to generate income went up astronomically. This is what a bubble is. When you have too many dollars chasing something, it drives the price up. The U.S. government, because it didn't have the constraint of the gold standard, could print as much money as it wanted. And it did.

The money supply, which was about 1 trillion dollars in 1975, is now estimated to be closer to 12 trillion dollars. This is a staggering increase, but there was nothing stopping the government from creating this money. It had no credit limit.

From 1987 to 2006 Alan Greenspan was the head of the U.S. Federal Reserve, or the "Fed" as it's referred to. The Fed is basically the central bank of the United States and was created to stop runs on banks. It now has a number of responsibilities, but one of the main ones is to manage the nation's money supply through monetary policy.

The Fed can use monetary policy to try and control the economy. If there is high unemployment the Fed might increase the money supply, giving people more money to spend, which is good for the economy. It might also lower interest rates, which inspires people to buy goods and homes on credit. This in turn encourages business to borrow money to purchase equipment and hire additional employees to operate that equipment.

Inflation is often a concern for governments because prices that go up too quickly can damage the value of the dollar. So the Fed can use monetary policy to raise interest rates to slow down an economy, or reduce the money supply. This means fewer dollars chasing goods and services, which hopefully causes those prices to fall, or at least to stop accelerating.

So while Alan Greenspan was head of the Fed, the money supply grew from less than 4 trillion dollars in 1987 to almost 12 trillion in 2006. Clearly Mr. Greenspan was a big fan of the printing press. He cranked it up into overdrive to flood the world with U.S. dollars. The economy was growing, but more and more of America's traditional industrial base, the companies that actually made stuff, were moving offshore to places like China. So at the same time as the world was being flooded with liquidity or dollars, the economy that backed those dollars was actually getting weaker.

What was happening was that the emperor had no clothes. With no gold behind the dollar and the U.S. economy losing its traditional strength, the value of the dollar was now questionable. But everyone was in the game. If countries started openly declaring that they felt the U.S. dollar had less value, it would adversely affect the value of their U.S. dollar holdings. So everyone just kept quiet and played along. And the Fed happily obliged and kept printing more dollars.

After the tech bubble burst in 2000 and foreigners became cautious about putting their dollars into the U.S. stock markets, they decided it would be better to own actual tangible assets, like houses and commercial properties and buildings. Real assets.

At the same time, Alan Greenspan had to figure out how to deal with the bursting of the tech bubble. He decided to inflate another bubble, this time in real estate. He dropped interest rates dramatically, which made borrowing money very cheap, and inspired many people to enter the home market. This caused the housing market to improve dramatically.

Soon, people who had never considered buying a home decided it was time to get in. Some people who perhaps shouldn't have been able to afford a house also got the housing bug. Now, as long as house prices were rising, this was okay. If you could get a mortgage and if the price of your home continued to increase, you could always tap into the equity to borrow money to help you pay the mortgage. As long as the value of the home went up, everything was fine.

Capitalists, being great spotters of trends, noticed this bubble inflating and stepped into the fray. They realized that they could get someone a mortgage who probably wouldn't have qualified before. These were called "NINJA" mortgages–"No Income, No Job, No Assets." No problem, you could get a mortgage anyway. With so many new homebuyers entering the market, prices couldn't help but go up. These mortgages were called "sub-prime," because these mortgages holders would not meet the traditional bank definition of a prime loan.

Historically, you got a mortgage from a local financial institution. The bank or loan company knew the city or area your home was in and took a very critical look at your ability to repay. It did not want to risk losing its investment, so it had high standards.

Those standards were regulated by the U.S. Government, which wanted banks to lend money responsibly. Yet with the worldwide trend towards "less government" and industry that was self-regulating, there was no one really policing the mortgage industry.

Then "Wall Street" entered the game. It realized it could take a bunch of mortgages and bundle them together and sell them around the world to other banks or hedge funds. These bundled mortgages, Collateralized Debt Obligations or CDOs, were sold to institutions that didn't look at them critically enough to realize what they were buying. The assumption was that since there was collateral—the value of the house—they were a good investment. As long as the house that the mortgage was on maintained its value, even if the mortgage holder defaulted the lender could always reclaim the house or collateral.

The activity in the housing market between 2001 and 2006 was frenzied. Builders couldn't build houses fast enough. People flipped houses they bought and renovated. People bought second homes and condos as investments. There seemed to be no stopping it. As long as the value of the homes kept going up, the party just kept rolling along.

Even Ben Bernacke, who took over the Fed after Alan Greenspan left in 2006, was on record as saying, "There has never been a time when the value of homes fell, across the nation." He meant that you may have a market where an industry is having a hard time and the house prices fall in that town or city, but never have house prices fallen simultaneously across the U.S. Well, that was until 2007 when the bubble burst, and that's exactly what happened. And they started falling hard.

Many people who had purchased homes that were beyond their means had been able to take home equity loans and use that money to help pay their mortgages. As long as the value of the home was going up, and therefore your equity was increasing, this worked fine. But what happened when the value of the home stopped going up and actually started dropping? People were having problems paying their mortgages, and eventually many people had to walk away from their homes.

The creative financial wizards in the U.S. had also come up with some really creative products to help entice people into homes who perhaps couldn't afford one. "Interest-only mortgages" or "Adjustable Rate Mortgages" had relatively small payments up front for the first couple of years. Then the mortgage would reset at a higher rate. These "sub-prime" mortgages helped convince lots of people to get into the market, but then just as the housing bubble burst and home values started dropping people's mortgages began resetting at the higher rates. It was a double whammy. People who had been paying $700 with an interest-only mortgage now had to pay double or even triple that. If the value of their house had continued to rise, they might have had a shot at using their home equity loan to help with the increase, but with the value of their home actually dropping this wasn't an option.

And as people started walking away from mortgages and the value of homes was falling, the banks and institutions that had purchased some of these sub-prime mortgages started to realize that there was a major problem. Credit agencies that rate such products had failed to properly assess the quality of the loans. The Wall Street banks that had sold the products weren't upfront about the quality of the loans. Now the jig was up, and the world experienced the "sub-prime" meltdown of August of 2007. This is where banks and financial institutions all over the world had to come clean to their shareholders about the problems with these products, and many took huge losses as a result.

Now the problem is a self-reinforcing one. As house prices fall, more and more Americans are having trouble paying their mortgages. As they default on their loans, and the banks foreclose on their homes, and cities become full of homes being sold in distress, it drives the home prices down further. And many of the people who made a living building new homes are now out of work and having trouble making their payments.

So here we are. Millions of Americans are "upside down" or "underwater" on their mortgages, meaning that their mortgage is actually worth more than the value of their home. The housing market is collapsing, Americans are losing their homes in record numbers, banks are in trouble because of their shaky loans, and economies are struggling because the credit markets have tightened up. With so much bad debt out there, no one wants to lend money, and credit is what drives a capitalist economy.

The U.S. Government is spending trillions of dollars trying to save financial institutions and stem the financial bleeding. And now the sickness that started in the U.S. is spreading around the world with lethal speed. People throughout the world are watching their retirement funds drop precipitously, their home values plummet, their jobs disappear, their lives turn upside down after such a long period of seemingly endless growth.

It didn't have to be this way. Governments could have done their job. They could have supervised the financial markets to make sure the shenanigans weren't happening. They could have had tighter controls over the printing press to make sure the money supply didn't expand so rapidly, and they could have raised interest rates and made borrowing money more difficult. Many of these policies are the work of Alan Greenspan. He chose this path, and now we are paying the price for it.

He chose to try and ignore the natural cycles of expansion and contraction that economies go through and attempted to have an ever-expanding economy. It has collapsed under the weight of itself. Some say the job of the Chairman of the Federal Reserve is to remove the punch bowl just when the party gets going. Alan Greenspan should have raised interest rates to avoid the housing bubble, but he seemed to ignore it and not take responsible action.

In short, governments could have behaved like households are supposed to. You only have so much income and so many assets, and you have to try and live within those means. If you overextend yourself, run your credit cards too high, put too many luxuries on your line of credit, take out too many second and third mortgages, sooner or later it will catch up with you. Either the bank will cut off your credit or, if your income doesn't allow you to cover your debts, you'll have to declare bankruptcy.

In an article in the *Financial Times* entitled *Welcome to a World of Diminished Expectations*, Willem Buiter wrote:

> Capitalist market economies are inherently cyclical. The private credit system is intrinsically prone to alternating bouts of irrational euphoria and unwarranted depression. Busts play an essential role. They clean up the mess created during the boom by inflated expectations, over-optimistic plans and unrealistic ventures. These become embodied in unsustainable household debt, productive capacity with no foreseeable use, excessive corporate and financial sector leverage and enterprises whose only asset is hope. The correction is painful, even brutal: unemployment rises, as do defaults, repossessions and bankruptcies. We entered such a cathartic phase around the turn of the year in both the U.S. and the UK. Continental Europe is not far behind.[1]

"It feels like I'm at the epicenter of the biggest financial crisis in history," says Meredith Whitney, the Oppenheimer & Co. analyst renowned for her smackdowns of America's biggest banks.

Many are calling what we're experiencing the second great depression. Many suggest it's not that bad and say you don't see long lines of people waiting for food, yet like so many of the modern systems we've created, we've merely moved people from breadlines into grocery stores with food stamps. According to the U.S. Department of Agriculture, 32 million or 1 out of every 10 Americans receives foodstamps as of April 2009. This is happening while food banks across North America are struggling to keep food on their shelves for the growing numbers of people now forced to use their services. We have simply shifted the food lines inside and out of sight. Plus the unemployment rate in the U.S. is exploding, taking a huge toll on families. And finally, the cat is out of the bag on the U.S. Debt. The debt, which has built up over years and is exploding with Bush and Obama stimulus plans, is close to $12 trillion, which is an incomprehensibly high number. But if you include "unfunded liabilities" it is closer to $45 trillion. These are the obligations that the government has for Medicare and Social Security, which millions of American baby boomers are about to start claiming. These are obligations that the U.S. Government has but which it has not set money aside for; hence they are "unfunded." The U.S. Government has not balanced its yearly budget since Bill Clinton was in the White House, and it now spends more than a trillion dollars more than it takes in through tax revenues. And now it has these huge new obligations to spend more money that it simply doesn't have. With the economy in free fall, it has no real hope of finding the money to pay for them either.

It can simply print more dollar bills, but that assumes that someone will accept those IOUs and lend the government the money. As its debt skyrockets it will become increasingly difficult for the U.S. Government to finance that debt, and it will have to pay higher interest rates to attract the money.

And finally, there is the possibility that the U.S. Government will simply have to default on its loan obligations. It will have to basically declare bankruptcy. Countries around the world are becoming increasingly concerned about the U.S. level of debt and starting to rethink a global reserve system based on the U.S. dollar. Suddenly central banks are diversifying their holdings, retaining a lower percentage of U.S. dollars and increasing their Euro and gold holdings. This does not bode well for the U.S. Sooner or later, the world will no longer accept this system, and the U.S. will be left a far poorer nation. It's a terrible thought, but as long as it continues to increase its spending without being able to offer a solution to how it's ever going to pay off its debt, a default is no longer unthinkable.

What that means to you is that you simply cannot continue to think that you can count on your government to bail you out. It cannot bail itself out, so stop thinking it will be able to help you. For a while it will be able to continue to fund unemployment payments and food stamps and some of the programs that have been a stopgap measure for people falling on hard times, but eventually this will have to end.

The message here is that sooner or later you're going to be on your own. I hate to be the one to say this, but there really doesn't appear to be a way out. Economists like to suggest that as long as the economy continues to grow it's all right for the government debt to grow. Eventually the continuously expanding revenue will pay off the debt from the ever-expanding economy, and that will solve the problem. It appears that this doesn't hold true anymore. The debt is too large and it's questionable whether it will ever be paid off unless the government defaults.

The one thing that is clear is that the problem is enormous, the government is not addressing it, and you need to reduce your expectations about what it can do for you. It's part of the American spirit to be independent; now we're just going to have to be more independent than at any time in most of our lives.

On May 28, 2008 Richard W. Fisher, President and CEO of the Dallas Federal Reserve Bank, estimated the obligations of the U.S. to be actually $99.2 trillion. And if we add the new debt, courtesy of Wall Street bankers, the obligations of the U.S. taxpayer rise to an impossible-to-repay sum of **$105,200,000,000,000.00** ($105 Trillion).[2]

Pretty soon, something is going to give.

4   Peak Oil

People living in the western world have a pretty amazing standard of living. We owe a huge debt to cheap and abundant fossil fuels for providing us with this standard of living. Nature provided us with this incredible one-time gift of fossil fuels when eons ago plant and animal material died and through a complex combination of heat and pressure and time we ended up with coal and oil and natural gas.

When we dig up this hidden genie and let it out of the bottle, it can accomplish incredible things for us. It can displace huge amounts of human effort to do tasks we once had to do through back-breaking personal effort. Coal was the first fuel we discovered to replace wood as our primary source of heat. It was pretty dirty when it burned, so when we discovered we could burn oil and natural gas that we were able to drill and extract from the ground, we started to replace coal, at least to heat our homes.

In most of our North American homes we heat with natural gas or home heating oil, which are relatively clean burning and can be supplied to our homes by someone else, either by truck or pipeline. When I say "relatively" clean burning, I mean we have built systems like furnaces that burn the fuel out of sight and send the waste materials and gases out the chimney to the atmosphere. While natural gas burns "cleaner" than home heating oil, it still releases pollutants and, worst of all, the climate-warming gas carbon dioxide.

When these fossil fuels were being created they were storing carbon in very concentrated forms. When we burn hydrocarbon-based fossil fuels we release this carbon back to the atmosphere, and it's this build-up of carbon dioxide (CO_2) that is causing the climate to warm. So even though many assume that natural gas is a "clean" way to heat, it is still releasing stored carbon into the atmosphere.

Much of the oil we extract from the ground is used for transportation. We use it to power our planes, trains, and automobiles. It also powers farm tractors and industrial machines that make our life easier. The potential energy in a barrel of oil is incredible. Three tablespoons of crude oil represent the equivalent of eight hours of human manual labor. Every time you fill up your gas tank your car is using the equivalent of two years' worth of human labor. It is estimated that if you took the potential in a human's entire life of manual labor, it could be replaced by three barrels of crude oil. How discouraging is that! Your entire life of digging and hoeing and cutting and chopping, displaced by just three barrels of crude. Discouraging—and scary, if we were to ever run out of oil.

The potential of this resource is massive. When you take a barrel of crude oil and process it at a refinery and break it down into its various parts, you get a number of products like gas and diesel fuel for cars and trucks, jet fuel for planes, and a series of other components like liquid petroleum gas, propane, and asphalt. Each of these products has huge potential in it. Think of the energy required to get a huge jet airplane into the air. Think of the energy needed to move one of those 18-wheel trucks you see on the highway, especially when it's fully loaded with goods. And think of gasoline moving around your 3,000-pound car. Try the "Push" test. Find an abandoned parking lot. Put your car in neutral, have someone behind the wheel steer, and try pushing your vehicle. See how far you can push it before you get tired. I'll bet it won't be very long. Now consider how much energy is required to get that vehicle up to 60 miles per hour (100 kilometers per hour) on the highway and keep it there. It's enormous. The energy density of gasoline and fossil fuels is immense. These relatively lightweight and portable fuels have given us incredible mobility and allowed us to be freed from so much of the mundane work of daily life and therefore able to pursue other interests.

So think of the challenges we would encounter if we found ourselves without abundant and cheap energy. Suddenly the world would be a very different place.

Like so many things on earth, oil is finite. There's only so much of it we can get at. In the 1950s a geologist named Marion King Hubbert who worked for Shell Oil made a prediction that the United States would hit "peak oil" in 1970. At the time the U.S.US was the largest producer of crude oil in the world, and Hubbert's peers mocked him. But low and behold, when we started looking back in the mid '70s, it was apparent that the U.S.US did indeed hit peak oil around 1970. There was a small

Peak Oil: US Oil Production

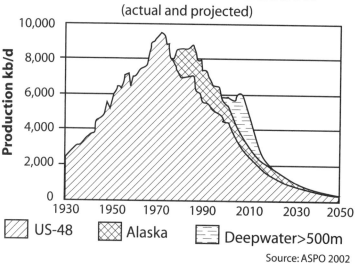

(actual and projected)

US-48   Alaska   Deepwater>500m

Source: ASPO 2002

uptick when oil was discovered in Alaska and again when it was discovered in the Gulf of Mexico, but this did not change the fact that in 1970 the U.S.US produced the maximum amount of oil it would ever produce. It didn't suddenly run out of oil; it just couldn't produce more, no matter how hard it tried, and oil supplies went into decline. Peak oil is essentially the point where you have removed half the oil. The challenge is that the half that remains is much harder to get out of the ground. It may not be economical to get at some of it because it will be too expensive.

Hubbert predicted the U.S. would hit peak oil in 1970 and he also made the prediction that the world would hit peak oil right around now. There is some disagreement about whether we have hit it, or will hit it soon, or whether it's very far in the future, but the reality is that it looks as if the world has hit peak oil. Since 2005 the world has produced about 85 million barrels of oil a day. This happened as oil prices rose significantly. If you had a million barrels of oil in the ground, wouldn't this rapidly increasing price make you want to bring it to market? That didn't happen, and there's a good chance it's because there just wasn't any more readily available oil to pump.

There are a growing number of geologists who are now suggesting that we have passed or are very close to hitting peak. The main critics of the theory of peak oil tend to be economists who feel that with any commodity if the price is high enough new supply will always enter the

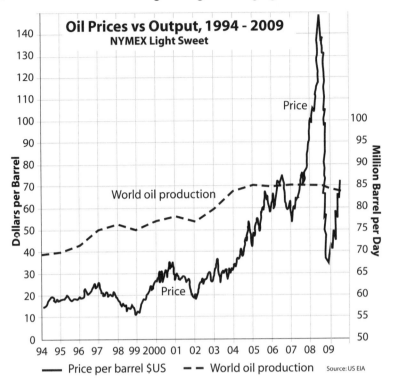

market to fill the demand. Economists don't seem to understand the physical limitations on how much oil lies under the planet's land masses and oceans, but geologists do.

A noted voice is Matthew Simmons, who is a Wall Street energy investment banker who worked as an energy advisor to President George W. Bush. His book *Twilight in the Desert: The Coming Saudi Oil Shock and the World Economy* is a landmark in the exploration of peak oil. Saudi Arabia is the largest producer and exporter and claims to have the largest reserves of high quality crude oil on the planet. Simmons suggests that there's every indication that Saudi Arabia has hit peak oil, and if they have then the world has as well. Proving this is a difficult challenge. Saudi Aramco is the state oil company of Saudi Arabia and therefore does not have to release information to the public. Publicly traded oil companies have to make information on reserves and production public.

Saudi Arabia is one of 12 members of the Organization of Petroleum Exporting Countries (OPEC), which represents the largest exporters of oil in the world. In the mid '80s these countries made a rule as a cartel that their production would be based on their stated reserves. So the amount of oil you could sell and the resulting income were dependent on what you said your reserves were. In the 1980s OPEC's total reserves magically increased from 350 billion barrels to over 600 billion barrels without a single major discovery. With such a large percentage of the world's reported reserves belonging to these countries, this is a huge concern.

Saudi Arabia nationalized its oil company Saudi Aramco in 1980. Up until that time it was partly owned by publicly traded American oil companies who by law had to state what they honestly believed the oil reserves of the country to be. The company geologists at the time said it was 110 billion barrels. You'll remember a few years ago when Shell had to "write down" their proven reserves because they had been overly optimistic about how much oil they could get out of the ground. So, if anything, publicly traded oil companies try and underestimate the potential oil they'll get out of the ground in case they can't actually extract as much as they hope.

In the late 80's Saudi Arabia increased its reserves from 110 billion barrels to 260 billion barrels of oil without reporting a single new major discovery. Even though it claims to have exported about 46 billion barrels in the last 17 years, it still says its reserves are 260 billion barrels. Since oil is vital to the world economy it is quite disturbing that we can't rely on the information provided by the biggest players in the game. As long as their ability to pump oil is based on their stated reserves they have every incentive to fudge their numbers.

Since the math didn't add up it got Simmons wondering what their reserves really were. He couldn't get them to actually tell him what they really had, so he began a painstaking process of analyzing the reports that Saudi geologists and petroleum engineers presented at conferences. Professional organizations often share challenges they are experiencing to learn from each other and to move the industry forward. In looking at the type of problem Saudi engineers were having with their wells it was very clear to Matt Simmons that they did not have the reserves they claimed. The bulk of the oil they pumped came from a small number of giant and super-giant oil fields that had been producing for 40 years. If you look at the amount of water that was coming out of the wells and the amount of gas and seawater they had to pump in to maintain the pressure of the wells, it was clear that these wells were past their prime.[1]

It makes sense that at a certain point the ability to keep pumping more of this black liquid from the ground is going to hit some threshold or limitation. At 86 million barrels of oil a day, that's 1,000 barrels a second. It's enough oil to fill 5,500 Olympic-size swimming pools a day. It's a mind-boggling number. Surely at some point there won't be enough oil to keep this up and we'll go into decline.

This is the concept of peak oil, that we will hit a point where we will never be able to surpass current production. We won't suddenly run out of oil, we'll just have less and less to use each year. This has been transpiring while demand for oil continues to increase. Rapidly growing economies in India and China have been increasing their use of oil as they expand.

So we may end up with demand for oil increasing just as we see production falling. This is a huge problem and one governments around the world seem hesitant to discuss. Agencies like the International Energy Agency (IEA), which reports on the state of the world in terms of energy for the OECD countries, have for many years had a "don't worry be happy" attitude to official reserve data.

Each year the IEA releases its *"Outlook"* yearly summary. Its 2008 report marked a turning point in its approach to challenges facing the industry. The report spends a great deal of time discussing global warming and how the burning of fossil fuels is contributing to it. It also talks about world oil production hitting a "plateau," as it calls it, a point at which production will no longer increase. This is a fancy name for peak oil. Many experts would disagree with the timing, but at least the IEA is no longer dismissing those who say that peak oil is imminent.

One of the most significant items of the report is this:

> We estimate that the average production-weighted observed decline rate worldwide is currently 6.7% for fields that have passed their production peak. In our Reference Scenario, this rate increases to 8.6% in 2030. The current figure is derived from our analysis of production at 800 fields, including all 54 super-giants (holding more than 5 billion barrels) in production today.[2]

This is a staggering jump from the previous year, when the IEA suggested that the decline rate would be about 3%. It's an absolutely terrifying number because it indicates that the rate at which oil production is declining is very fast. The IEA suggests that new oil will be discovered, enough to not only make up for this depletion but also meet the increased demand from developing economies. People have made this assumption for many years now. But we are no longer discovering the large "elephant" oil fields, as they're called.

As George Monbiot notes:

> The world's problem is as follows. We now consume six barrels of oil for every new barrel we discover. Major oil finds (of over 500m barrels) peaked in 1964. In 2000, there were 13 such discoveries, in 2001 six, in 2002 two and in 2003 none. Three major new projects will come on-stream in 2007 and three in 2008. For the following years, none have yet been scheduled.[3]

Many prominent petroleum geologists believe we have hit peak oil. Kenneth Deffeyes in his book *Beyond Oil, The View from Hubberts Peak* projects that the world hit peak oil in November 2005. Matthew Simmons, author of *Twilight in the Desert* agrees with the 2005 date. Texas oil billionaire T. Boone Pickens made his fortune in the oil business. He claims we hit peak in 2006 and he has spent millions of his own dollars promoting his "Pickens Plan" (www.pickensplan.com) because he's tired of seeing $700 billion a year leaving the U.S. to go to foreign oil sources. His plan promotes massive investment in wind power to free up natural gas that is currently being used to generate electricity and switching that natural gas to transportation to stop the flow of dollars out of the country.

Colin Campbell and other noted geologists formed The Association for the Study of Peak Oil & Gas (www.peakoil.net), and they are trying to raise the world's awareness of this looming challenge. Of the 65 largest oil-producing countries in the world, up to 54 have passed their peak of production and are now in decline, including the U.S. in 1970, Indonesia in 1997, Australia in 2000, the North Sea in 2001, and Mexico in 2004.

While it's nice to see someone like T. Boone Pickens at least offering a plan, using natural gas for transportation may not be the answer. Natural gas is becoming more and more valuable as it becomes harder and harder to find. As early as 2001 a number of prominent energy organizations were still optimistic about the availability of natural gas in the U.S. The National Petroleum Council, Cambridge Energy Research Associates, and the US Department of Energy all projected that the production of natural gas would continue to grow to meet increasing demand up until 2020.[4] By 2005 though, their outlook had changed dramatically. Despite rising prices and increased drilling activity, domestic production was now in permanent decline. This may be due to the "natural gas falling off a cliff" syndrome. With an oil well, as the well is depleted you can continue to extract oil in ever-decreasing amounts over time. Finding a natural gas deposit is more like puncturing a bike tire. Gas will flow out very quickly, and then abruptly stop.

Natural gas is in serious decline in North America. As Kenneth Deffeyes reports in *Beyond Oil: The View from Hubbert's Peak*: "Between 1980 and 2002 the best of the natural gas targets were drilled. We're now being served the leftovers."

One plan to mitigate the precipitous fall in U.S. natural gas is through the use of Liquid Natural Gas (LNG). This requires asking countries like Russia and Iran with extra natural gas to use massive amounts of energy to refrigerate it to 160°C below zero which turns it from a gas into a liquid and also minimizes its volume. The LNG is loaded into boats and shipped to the location that needs it. At the unloading terminal lots of energy is required to turn it from a liquid back into gas so that it can be pumped into the natural gas pipeline.

LNG is very volatile. A stray spark of static electricity or a well-aimed rocket-propelled grenade could set off an explosion many times larger than a nuclear blast, as happened in Algeria in 2004, when 26 workers were killed and hundreds of others injured. The Rocky Mountain Institute in its report *Brittle Power: Energy Strategy for National Security* estimates that the energy content of one standard liquid natural gas tanker at 22 billion gallons of expanded gas is equivalent to .7 megatons of TNT (that's 1.4 billion pounds of dynamite). To put it another way, the explosive force of one LNG tanker is equal, roughly, to 55 Hiroshima bombs. Not surprisingly, no one wants these unloading terminals near their homes. Although LNG offers one of the only hopes of continuing with a "business as usual" approach to how we use energy in North America, in reality it isn't much of a hope. You simply can't import enough LNG to make up for the rate at which our natural gas supplies are declining.

In 2005 the U.S. Government undertook a study of peak oil and its author Robert Hirsch warned that peak oil would have such a catastrophic impact on the world that we would need to have a plan to mitigate the negative impact 20 years before peaking.[5] If peak oil were 20 years in the future, governments would need to start now. But governments have been using the IEA reports, which until 2008 were calling those who spoke of peak oil "doomsayers."

It was only in 2008, with the price of oil rising to $147/barrel and many pundits openly discussing peak oil that the IEA returned to the drawing board and issued a new report, which only raises the specter of oil field depletion in theoretical terms. The attitude of the IEA is that we can deal with the challenges if we just keep spending ever greater sums of money looking for more oil and gas. The problem with this theory is that the collapse of the world economies in 2008 has created havoc with the capital financing that these massive projects require. And not only has the financing dried up; the low price that oil has reached takes away the incentive to invest in a lot of this exploration. Even when high prices return, massive sums of capital have simply been wiped out, so it will make financing massive energy projects prohibitive.

There's a good chance we've already hit peak oil, and we have no mitigation plan. Governments have dropped the ball. Oil companies have been running advertising campaigns to convince us there's no real problem. We're like deer staring at the headlights as the freight train of peak oil lumbers toward us, and we can't seem to get out of the way.

Don't be like the smokers who chose to believe the pseudo scientists who were paid by tobacco companies for decades to tell us that cigarettes are perfectly safe. What the oil companies and governments aren't telling you CAN hurt you. It WILL have an incredibly negative effect on your life.

Governments may not be able to see reality, but you can, and section II gives you the tools you need to create your own "Plan B." Don't be a deer caught in the headlights.

5   Peak Food

There is a group called the "Club of Rome." It sounds like either a hip European dance club or a sinister, elite old boys' club. It is in fact a global think tank that published a report in 1972 called *Limits to Growth*. The report suggested that the world population, which had reached 3 billion in 1961, was at a theoretical threshold and that the planet couldn't support any more people without seriously damaging its ability to support life. The report sold 30 million copies and was a topic of conversation at the time, but the population continued to grow exponentially.

The theme was similar to a book published in 1968 by Paul Ehrlich called *The Population Bomb*, although this book was much more pessimistic in its suggestion that there would be large-scale starvation in the 1970s and 1980s. While there were definitely famines that took an enormous toll on populations in sub-Saharan Africa and other places around the globe, humanity was able to avert some of Ehrlich's worst predictions. Improvements in technology, the bioengineering of crops, and the eventual lowering of oil and natural gas prices, which allowed lots of natural gas to be converted into fertilizer, dramatically increased crop yields.

This book was similar to one written by Thomas Malthus in 1798 called *An Essay on the Principle of Population*. With Malthus' prediction of overpopulation causing poverty disproven time and time again, *The Population Bomb* was dismissed by many as another Malthusian catastrophe and ignored by many. Since we had averted disaster repeatedly, and since we could turn all those new humans into new consumers, growth had no limits and could continue indefinitely.

There are cracks starting to show in the theory of infinite growth though, as evidenced by the simple fact that the population of the planet doubled in just four decades, from three billion in 1960 to six billion

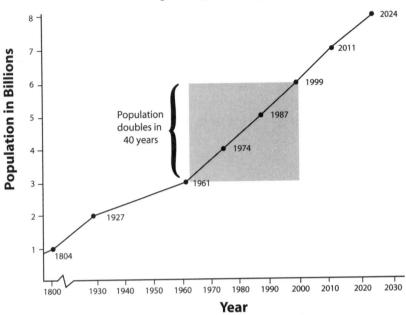

in the year 2000. That is a lot of people in a very short period of time. That's a lot of mouths to feed. If you look at the population of the planet over time, it grew very slowly for many centuries. Then we unlocked the fossil fuel genie, and suddenly farmers could grow huge surpluses of food to feed more mouths.

Basic logic tells us that this simply can't go on forever. Sooner or later the planet is going to run into some theoretical limit on how many people it can feed. When you look at how explosive population growth has been in the last 50 years, most people can make the leap and acknowledge that we've hit that limit or perhaps gone beyond it.

There's no doubt that during the period after *Limits to Growth* global governments were able to maintain the illusion that everything was fine and that limitless growth was sustainable. We also experienced a period of neo-conservatism where leaders like Margaret Thatcher and Ronald Reagan were elected on platforms of reduced government interference in the economy. The trend that they began is starting to show signs of being unmanageable.

In North America, interventionist governments had created systems to try and help farmers maintain their incomes as the price of the commodities (food) they grew went up and down with the market. Sometimes

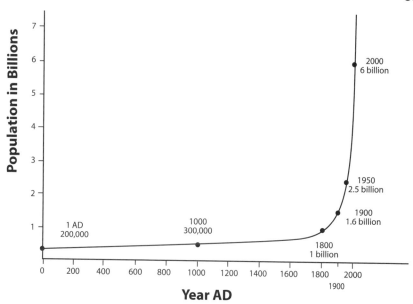

in periods of low prices governments would buy excess grain and store it for the future. This prevented the price from dropping too low. Then in a subsequent year when the price of grain was much higher, they would sell some of that excess into the market, reducing the price and regulating the gyrations that the price was prone to. While the farmers preferred the higher prices during the times when grain was in lower supply, they also realized that these programs isolated them from the worst of the fluctuations.

Needless to say neoconservative governments were going to have no part in this sort of intervention, and they began to gradually extract themselves from these sorts of programs. Larry Matlack, President of the American Agriculture Movement (AAM) has raised concerns that the U.S. no longer has a suitable inventory of grain on reserve. According to the United States Department of Agriculture's (USDA) Commodity Credit Corporation (CCC), as of May 2008 there is an inventory of 2.7 million bushels of wheat in inventory. This is enough wheat to make half a loaf bread for each of the 300 million people in America.

As Matlack says, "Our concern is that the U.S. has nothing else in our emergency food pantry. There is no cheese, no butter, no dry milk powder, no grains or anything else left in reserve." This is particularly troubling when you realize that these supplies of grain are often used for humanitarian relief. The images of starving people you often see in the news, with trucks full of bags of grain being delivered, are premised on there being grain somewhere to offer.

In the past, governments had years' worth of grain in storage to help them through rough times like prolonged droughts. But we've been drawing down that balance and have eaten through about two-thirds of the world's grain reserve in the last five years, and globally we have less than two months left. World rice reserves have plunged to 9 weeks' worth of consumption from 19 weeks as recently as 2001.

What's been happening is basically a perfect storm in terms of the food supply of the planet. As the population has been exploding, we've been pushing the land to its limits through fossil-fuel-intensive-agriculture. Whether it be natural gas for fertilizer, oil for pesticides and herbicides, or diesel for ever larger tractors, we have become completely dependent on cheap and abundant energy to feed us all. And it looks as if that free ride may be over with peak oil.

North Americans basically eat fossil fuels. For every 1 calorie of food energy we eat, 10 calories of fossil fuel go into getting it to our plates. This includes the natural gas for fertilizer, petroleum products for pesticides and herbicides, diesel for the tractor to plant the seeds, diesel for the tractor to apply chemicals during its growth, diesel to harvest it, diesel to truck it to a processing plant, energy at the plant to pump water to clean it or to boil water for cleaning or canning it, energy for the glass or steel for the container that the food ends up in, cardboard for the box they are packed in, diesel for the truck to get it to the chain store's central warehouse, energy to keep that building cold, diesel to take it from the warehouse to the food store, electricity to keep the lights on and refrigerators cold at the store, and finally gasoline for you to drive to the store and buy the food.

And if you didn't remember your own bags, don't forget the petroleum for the 4 trillion plastic bags produced from oil each year worldwide. The 14 billion plastic shopping bags given away in the U.S. each year cost retailers about $4 billion and use 12,000,000 barrels of oil. [1]

As early as 1974, a study led by John Steinhart, then at the University of Wisconsin at Madison, concluded that the U.S. food system had quadrupled its energy use between 1940 and 1970. It now takes between 10 and 15 calories of energy to deliver one calorie of food to a U.S. consumer. A head of lettuce, for instance, requires 2,200 calories of energy to produce when it's grown in California and eaten in New York, yet it provides only 50 calories of energy. By contrast, subsistence societies use about four calories of energy to produce one calorie of food.[2]

There is simply no denying that we use an incredible amount of energy in the production of our food. While this has allowed North Americans the luxury of spending the smallest percentage of their incomes on food of anyone on the planet, it has left us incredibly vulnerable to the problems of declining energy stocks and rapidly increasing energy costs. Many people are trying to move towards a more local or *"Hundred Mile Diet,"* but the reality for most North Americans is that we still have food that travels an average of 2,500 miles to get to our plates. That is going to change or become prohibitively expensive for many of us.

Climate change adds more fuel to the fire. A one-degree increase in average temperature on the planet will have a hugely negative effect on global food production. It will be most pronounced in countries nearest the equator. This poses a challenge because in the past governments have been able to rely on trade to make up for deficiencies in their own production.

In the spring of 2008 we got the first glimpse of the future of food in a warming and fossil-fuel-constrained future. Prices rose dramatically and countries began hoarding food that they traditionally exported to other countries. World Vision Australia head Tim Costello called the situation desperate and chronic. "It is an apocalyptic warning," he said. "Until recently we had plenty of food. The question was distribution. The truth is because of rising oil prices, global warming and the loss of arable land, all countries that can produce food now desperately need to produce more."[3]

Lester Brown, president of the World Policy Institute states:

> This troubling situation is unlike any the world has faced before. The challenge is not simply to deal with a temporary rise in grain prices, as in the past, but rather to quickly alter those trends whose cumulative effects collectively threaten the food security that is a hallmark of civilization. If food security cannot be restored quickly, social unrest and political instability will spread and the number of failing states will likely increase dramatically, threatening the very stability of civilization itself.[4]

Lester Brown, in his book *Who Will Feed China: Wake-Up Call for a Small Planet*, noted that when Japan, a nation of just 125 million, began to import food, world grain markets rejoiced. But when China, a market ten times bigger, starts importing, there may not be enough grain in the world to meet the need and food prices will rise steeply for everyone. Analysts foresaw that the recent doubling of income for China's 1.2 billion consumers would increase food demand, especially for meat, eggs, and beer.

In an integrated world economy, China's rising food prices have become the world's rising food prices. China's land scarcity has become everyone's land scarcity. And water scarcity in China will affect the entire world. China's dependence on massive imports, like the collapse of the world's fisheries, will be a wake-up call that we are colliding with the earth's capacity to feed us. Brown suggests, "It could well lead us to redefine national security away from military preparedness and toward maintaining adequate food supplies."

Governments around the world seemed to be caught completely off guard and had few quick and easy solutions to the problems. *The Economist* magazine claimed in an article entitled "The End of Cheap Food" that their food-price-index was higher than at any time since it was created in 1845. Wasn't that about when Robert Malthus first started warning about this?

Food riots broke out throughout the world in the spring of 2008. Shortages of essentials like rice and flour have a much more profound effect on poor people. According to the USDA, Americans spend less than 10% of their disposable income on food; people in many developing countries can spend upwards of 90% of their income on this essential. If the price of food basics begins to rise rapidly and their incomes don't keep pace, it makes for a very dangerous situation.

In developing economies like China, as more people earn more income there is a negative effect on the amount of food available in the world. In 1985 the Chinese consumer ate 44 lb (18 kg) of meat a year but in 2008 ate over 110 lbs (50 kg). This pushes up demand for grain because it takes 12 lbs (5.4 kg) of grain to produce 1 pound of beef. At the same time, grain subsidies in the US are diverting corn from food into the biofuel ethanol.

It was always the belief of governments that countries should specialize in whatever they did best. Some countries might make cars, and others might grow grain, and if they traded, each would be better off. But food self-sufficiency is a priority for a number of governments, and as food shortages began to occur in the spring of 2008 countries like Thailand that used to export rice stopped and began keeping that excess for their own population. India, Vietnam, China and 11 other countries limited or banned exports of rice, and Pakistan, Bolivia and 15 other countries stopped exporting wheat. Halting these exports only increased the severity of the problem.

U.S. trade representative Susan Schwab notes, "One country's act to promote food security is another country's food insecurity." The country that used to receive that food now has a problem. "If every country in the world decided it wanted to produce its own food for consumption," Ms. Schwab says, "there would be less food in the world, and more people would be hungry."[5]

In the future, it's quite conceivable that there will be less trade in food as governments attempt to make themselves more food secure. In the past, globalization has often prevented farmers from being able to make essentials for their local populations. With heavy subsidies and economies of scale, American farmers could grow corn much more cheaply than small Mexican farmers, but when the U.S. pulled back exports of corn when more and more of it was diverted to ethanol, Mexicans suddenly found themselves short of corn for tortillas, a staple of their diet, and many took to the streets to protest.

Governments around the world are now having their militaries develop scenarios for the outcome of global climate change, and many paint a very unpleasant picture of food in the near future. It appears that the warming climate, reduced rainfalls, rivers and aquifers with less water for irrigation, and a still-growing world population all point to a huge challenge. Australia has seen the first glimpse of this with a drought that has lasted for years and has limited its ability to ship any wheat abroad.

Countries that aren't growing enough food for their own population are going to have problems importing it from abroad. Even if they have the money, there may not be food available.

If they can't feed their populations, starving refugees will flood across borders and it will be difficult for governments to keep a lid on social order.

A final item to contemplate is how our food system, like so many of our industrial systems, has become a "just-in-time" process. Car makers found that they saved money if, rather than having a massive inventory of every part a car needed in its assembly, suppliers delivered the components just-in-time to put right into the car. They spent more on jets to rush crucial parts when the system broke down, but overall this saved them money. Many modern systems, such as our food system, have followed this model. The problem is that if inputs don't keep constantly flowing into the system, the disruption has a very negative consequence for anyone dependent on it, which is most of the population.

In the spring of 2008, as oil prices skyrocketed and the price of diesel fuel soared, truck drivers in Britain threatened to strike. *London's Daily Mail* newspaper depicted the possible results in an article entitled *"Nine Meals from Anarchy: How Britain Is Facing a Very Real Food Crisis."* The article claims it would take just nine meals or three days without food on supermarket shelves before law and order started to break down and British streets descended into chaos. [6]

One of the least publicized aspects of the attacks of 9/11 in New York City was exactly what the Daily Mail was talking about. In Manhattan, within days of the attacks store shelves were becoming bare and restaurants were running out of food. Many of the bridges and tunnels in and out of Manhattan were closed and trucks could not deliver food to the city. Someone with a well-stocked pantry might not have noticed the disruption, but someone who foraged for food daily without planning ahead would have found the situation quickly untenable. Having a significant stock of food doesn't look like such a bad idea in this context.

So one thing is for sure. You will be spending a much higher percentage of your income on food. We may not get to the point where it's 90%, but it will be higher, and much will depend on how quickly we run out of affordable fossil fuels and how quickly climate affects the ability of farmers to grow food. Your diet will become increasingly local and many of the items we now regard as commonplace, like pineapples, or January strawberries in the north, will become luxuries once more, out of reach for many of us. We have made the mistake of taking these items for granted, but they were always a luxury, and we need to treat them as such.

The greatest way to alleviate this challenge is to become proficient at growing your own food. There's never been a better time to pick up a shovel and a package of seed. Chapter 14 will get you started on the right foot.

6   Peak Water

All forms of life need water. Even single-celled animals and plants like bacteria and algae need water to survive and function. Living organisms are made up of approximately 60 to 70% water by weight. Water covers about 75% of the surface of the earth. The amount of water on the earth remains fairly constant, as it is continually "recycled" through evaporation and precipitation. This cycling of water plays a role in regulating the climate of the earth.

Less than 3% of the earth's water is fresh water and most of this fresh water is frozen in the ice caps and glaciers. Fresh water is a critical part of our world that is in relatively short supply!

The International Energy Agency notes:

> The Earth has a finite supply of fresh water, stored in aquifers, surface waters and the atmosphere. Sometimes oceans are mistaken for available water, but the amount of energy needed to convert saline water to potable water is prohibitive today, explaining why only a very small fraction of the world's water supply derives from desalination.[1]

The head of the Food and Agricultural Organization of the United Nations (FAO) says two-thirds of the world's population could be threatened by water shortages by 2025. The World Bank reports that 80 countries now have water shortages that threaten health and economies while 40% of the world's population—more than 2 billion people—has no access to clean water or sanitation. In this context, we cannot expect water conflicts to always be amenably resolved.

Fortune magazine recently said, "Water promises to be to the 21st century what oil was to the 20th century: the precious commodity that determines the wealth of nations." Water, like oil, is going to become increasingly important on the world stage. In North America we're already seeing water shortages starting to have a negative impact on many people's standard of living.

One month after becoming the Obama administration's new Secretary of Energy, Nobel prize-winning-scientist Steven Chu warned that California which currently produces nearly a third of the country's food supply could face serious challenges because of water shortages brought on by climate change.

"I don't think the American public has gripped in its gut what could happen," he said. "We're looking at a scenario where there's no more agriculture in California." And, he added, "I don't actually see how they can keep their cities going" either.[2]

California's Department of Water Resources projects that it will deliver just 15% of the amount that local water agencies throughout California request every year. In 2007 Governor Arnold Schwarzenegger declared drought emergencies in a number of counties while wild fires that formerly happened during a contained fire season now burn all year long. Georgia's Governor Sonny Perdue declared a water supply emergency in north Georgia in 2007 as its water resources dwindled to a dangerously low level after months of drought. *Atlanta* magazine recently ran a cover photo of a glass half filled with water and entitled it "This Glass Is Half Empty." In the summer of 2007 the city was only months away from running out of drinking water through a combination of low reservoir levels, a prolonged drought, and disputes over water through its jurisdiction controlled by the US Army Corps of Engineers.

At the same time that water is becoming scarcer, multinational corporations like Vivendi, Suez, and Bechtel are trying to increase their control of water supplies throughout the planet.

In an article in *Mother Jones*, water expert Maude Barlow discusses factors leading to the commercialization of water.

> I think the reality of the scarcity and the pollution of the world's surface water has just suddenly become real to people. And whereas 20 years ago you couldn't imagine getting most of your water from bottles, it's just become an accepted part of people's lives now. On planes, in restaurants, everybody drinks bottled water; you carry it around in your pocket. So it came first from scarcity, people needing access to clean water in their lives. And then the view that it was an okay thing to start commodifying water and using it in this way. Of course, behind all this are the big corporations. They've been aggressively promoting and marketing their water as better, as cleaner, as purer, as safer—which it is not. And it is to their advantage to let the public systems of the world's

water deteriorate while they get to make huge amounts of money off people's need for clean water.[3]

Even before the world began experiencing water shortages, there were conflicts over the ownership of water in rivers that run through more than one country. In the 1970s there were tensions over the Tigris-Euphrates River between Iran, Iraq, and Syria, and now Turkey (where the river begins) has become involved. Pakistan uses the Indus River to irrigate a huge area to grow food, but the river's headwaters are in India. As the glaciers that feed river systems like this melt, there will be conflict over who gets to use the water, and India and Pakistan, which are both nuclear armed, have already had a number of declared and undeclared wars.

In China, with 1.3 billion people, if rainfall affects river flows, growing enough food for its population will become increasingly challenging.

Fred Pearce, in his book *When The Rivers Run Dry* looks at why the Indus River in Pakistan, the Yellow River in China, the Rio Grande and Colorado in the United States, the Amu Darya in Central Asia, the Nile in Egypt, and many others are seeing so much of their flow depleted. The main reason is food crop irrigation.

> We are using these rivers to death. And we are also pumping out underground water reserves almost everywhere in the world. With two-thirds of the water extracted from nature going to irrigate crops—a figure that rises above 90 percent in many arid countries—water shortages equal food shortages.
>
> Farm yields per hectare have been stagnating in many countries for a while now. The green revolution that caused yields to soar 20 years ago may be faltering. But the immediate trigger, according to most analysts, has been droughts, particularly in Australia, one of the world's largest grain exporters, but also in some other major suppliers, like Ukraine. Australia's wheat exports were 60 percent down last year; its rice exports were 90 percent down.[4]

It really appears as though the Club of Rome might have had it right back in the 1970s when it suggested that we should be trying to limit population growth.

American agronomist and environmentalist Lester Brown, in his book *Who Will Feed China?: Wake Up Call for a Small Planet*, predicted in 1995 that China would soon not be able to feed itself largely because demand is rising sharply at a time when every last drop of water in the north of the country, its major breadbasket, is already taken. The Yellow

River, which drains most of the region, now rarely reaches the sea except during the short monsoon season.

A quick look at the variety of books available on the water challenges we face gives a pretty good idea of the scope of the problem: *When the Rivers Run Dry: Water—The Defining Crisis of the Twenty-first Century* by Fred Pearce; *Cadillac Desert: The American West and Its Disappearing Water* by Marc Reisner; *Last Oasis: Facing Water Scarcity* by Sandra Postel; *Water Follies: Groundwater Pumping and the Fate of America's Fresh Waters* by Robert Jerome Glennon are just some of the titles available from a seemingly endless list.

The Great Lakes are a vast inland sea representing over one-fifth of all surface fresh water on the planet. More than 40 million Americans and Canadians draw their drinking water from the lakes, and they are critical to public health, commercial interests, and regulating the climate. But there are causes for concern, such as declining water levels and invasive species. As southern states begin to experience water shortages, there will be enormous pressure to divert some of this water.

The Ogallala Aquifer is a large underground water source located beneath the Great Plains in the United States. It covers approximately 174,000 mi² (450,000 km²) from South Dakota and Nebraska to New Mexico and Texas. Almost 30% of the irrigated land in the United States overlies this aquifer system, which provides 30% of the nation's ground water that is used for irrigation. It also provides drinking water to 80% of the people who live near it.

Underground water reservoirs need to be replenished over time by the water cycle. But much of that part of the U.S. is quite dry and rain often evaporates quickly before it is able to make its way to the aquifer. So more water is being taken out of the Ogallala Aquifer than is being put back in. In fact, some estimate it will be dry within 25 years. That will have a huge impact for the people who rely on it for drinking water and for the crops that depend on it for irrigation.

The Colorado River, which is of huge importance to so many in the Southwestern United States, is also in trouble. It has been used so extensively for human needs and irrigation that it no longer reaches the ocean, and researchers from the Scripps Institution of Oceanography warn that the river's reservoirs could dry up within 13 years.

Mexico City now shuts off water to millions of its 20 million residents for several days because water reserves in the lakes that supply it are at historically low levels. Los Angeles, meanwhile, is toying with water rationing to deal with its water shortages. It seems that water isn't an issue for just China and the Middle East.

Daniel Zimmer and Daniel Renault, in their 2003 study, determined that it took 264 gallons (1,000 L) of water to produce 2.2 lbs (1 kg) of wheat; 370 gallons (1,400 L) to produce 2.2 lbs of rice; and 3,434 gallons (13,000 L) to produce 2.2 lbs of beef.[5] As people in developing countries earn more income and develop a taste for a western diet, more water will be required to provide more animal protein.

As we find water in increasingly short supply it is going to radically affect our lives. You will need to change your relationship with water by not taking it for granted, and you need a plan to harvest and store it in the future. You can go weeks without food, but only days without water, so get ready to be serious about that precious, life sustaining liquid.

7 Climate Change

The climate is changing. I know you're tired of hearing about it. We've been hearing about it for years, and yet nothing too radical has happened to most of us. Nothing much, unless your house was flooded in hurricane Katrina, or you experienced the 500-year floods in the U.S. Midwest that left the city of Cedar Rapids, Iowa under water, or you live on an island like the Maldives, which look as though they won't exist in the fairly near future because they'll be under the Pacific Ocean as it rises.

It hasn't affected most of us too much. Many of us saw Al Gore's movie *An Inconvenient Truth* and were shocked and awed about how bad things really are. We saw how much carbon dioxide is in the atmosphere, and we decided we had to make some changes. We left with great resolve to change our ways and reduce our carbon footprint. But then one thing led to another, we had to keep driving to work, it was too hard to get a bus to that part of the city, and we had to visit the grandma on the other side of the country, and next thing you know, we were right back consuming non-renewable liquid carbon-based fuels just as we had before we saw the movie.

I understand. It's very difficult to reduce your impact on the planet when for so long we created a culture that demanded it. Our cities sprawled into the suburbs, and owning a car was pretty much mandatory to get around. Somehow airlines were able to drive the price of airfare down so low that it didn't make sense not to fly anymore. Yet the reality is that our collective action is having a negative effect and it's happening at a much faster rate than anyone expected.

The Intergovernmental Panel on Climate Change (IPCC) is a scientific group evaluating the risk of climate change caused by human activity. The latest IPCC report (2007) claims unequivocally that the climate is warming as a result of anthrogenic (human) greenhouse gas emissions.

Temperatures will rise between 1.1 and 6.4°C during this century and we will experience more frequent warms spells, heat waves, and heavy rainfall as well as an increase in droughts, hurricanes, cyclones, and extreme high tides. Sea levels will probably rise 7 to 23 inches (18 to 59 cm).

There is no dispute any longer. The planet is warming and we're responsible. Even the International Energy Agency, which has usually been fairly reserved in its observations about climate change, came out with some strong words in its World Energy Outlook 2008:

> Preventing catastrophic and irreversible damage to the global climate ultimately requires a major decarbonisation of the world energy sources. On current trends, energy-related emissions of carbon-dioxide (CO_2) and other greenhouse gases will rise inexorably, pushing up average global temperature by as much as 6°C in the long term. Strong, urgent action is needed to curb these trends. Households, businesses and motorists will have to change the way they use energy, while energy suppliers will need to invest in developing and commercialising low-carbon technologies.[1]

National Snow and Ice Data Center (NSIDC) scientists monitor and study Arctic sea ice year round, analyzing satellite data and seeking to understand the regional changes and complex feedbacks that we are seeing. Senior Scientist Mark Serreze said:

> The sea ice cover is in a downward spiral and may have passed the point of no return. As the years go by, we are losing more and more ice in summer, and growing back less and less ice in winter. We may well see an ice-free Arctic Ocean in summer within our lifetimes. The scientists agree that this could occur by 2030. The implications for global climate, as well as Arctic animals and people, are disturbing.[2]

Jay Gullege, a respected climate scientist at the Pew Centers for Global Climate Change, recently noted that the best scientific models have been consistently underestimating the scale and pace of the change, year after year. In other words, scientists are correctly predicting what the results of climate change will be; they are just being conservative and underestimating how quickly these changes will take place.

Al Gore's documentary *An Inconvenient Truth*, includes photograph after photograph of glaciers as they were decades ago and as they are today. Some are basically gone.

Closer to home, we saw a record hurricane season in 2005 with Hurricane Katrina being one of the most deadly and expensive to ever hit the United States. Based on current climate change models, these occurrences will become the norm in hurricane-prone areas.

Islands in the Pacific like the Maldives are already under threat of rising sea levels, as are countries like Bangladesh and Holland and low-lying cities like San Francisco, New York, Beijing, Shanghai, and Calcutta.

As rainfall becomes erratic many people will experience water shortages, and the threat of one country taking water from a river that another country relies on will make the world a more dangerous place.

In his book *Climate Wars*, Gwynne Dyer takes a critical look at how governments and their military are now getting prepared for climate change. Even the U.S. military, which was distracted by eight years of climate change denial by the Bush Government, now sees climate change as one of its greatest concerns and the most likely cause of instability in the future. Food will become a huge issue with the changing climate. Dyer discusses how the Indian think tank Integrated Research and Action for Development (IRAD) concluded that a rise of only 2°C in global average temperature would cut Indian food production by 25%. Dyer notes:

> Since India's one billion people grow just enough food to feed themselves now, that is the equivalent of around 250 million Indians with nothing to eat—and it is unlikely that they would be able to buy food from outside, since, in any global heating scenario, most regions of the world that now have a substantial food surplus will be hit hard as well. [3]

What's most frightening is that many climate scientists now are shifting their focus away from proving that climate change exists and instead are devising strategies to deal with it. Ideas like sending large-scale structures into space to act like window shades are being discussed. Seeding the oceans with iron particles to encourage the growth of plankton that absorb carbon dioxide. And putting sulfur particles high into the stratosphere to deflect some of the solar radiation back into space. This is a bad thing. It means that many scientists are so concerned about humanity's ability to reduce its production of greenhouse gases that they are turning their attention to dealing with the outcome.

There are few bright spots in the climate change story. Water will be scarcer, food will be more expensive and harder to grow, storms will be stronger, coastal areas will flood, and the temperature is going to rise. If it hasn't affected you yet, ultimately it will. The bright spot for you, the reader, is that everything I discuss in this book for increasing your independence in all facets of your life will ultimately reduce your contribution to this problem. Whether it's integrating renewable energy to increase your energy independence, growing your own food to increase your food independence, or getting off the consumer bandwagon to rein in spending and increase your financial independence, all will help you have a smaller carbon footprint. Good for you, good for your wallet, and good for the planet. That's an inspiring message!

8 Why It's Different This Time

In the middle of difficulty lies opportunity.
Albert Einstein

I suggested earlier that "the world has changed," and in many respects the world today is much better than it was in the 1930s during the last major depression. Our standard of living in the western world is much higher. Our life expectancy is longer. There are fewer people living in poverty and there are many government programs designed to try and maintain a minimum level of comfort for those at the lower end of the social scale. These programs are obviously important for a society with so much wealth to make sure that people don't fall by the wayside.

It becomes challenging if we actually create a class of people dependent on the government and we reach a time when the government doesn't have the money to help them.

In the depression of the 1930s there were virtually no government entitlement programs. There was no welfare. There was no unemployment insurance. There was no government-sponsored health care. There was no old age security or retirement benefits. People were simply more independent. People didn't expect the government to help them out when things got tough because it never had in the past. When things got tough, people had to get going to get themselves out of their situation.

In developed western economies with so much wealth, no one would argue that it's fair to discard a segment of the population because its members are unable to find work. Any economy will always have a certain amount of frictional unemployment as people lose jobs and take time to find new ones. Economists think this is a good thing because it helps regulate wages. If there are too many jobs and low unemployment wages are high because employers have to compete to maintain employees.

Frictional unemployment is what unemployment insurance is for, to help tide workers over as they search for new work.

Economies also have some structural unemployment, where they simply cannot employ every member of society who wishes to have gainful employment. Structural unemployment also refers to what happens as economies change and new technologies displace old ones. Much of our steel is now produced in developing economies, so steel workers have had to retrain.

To deal with this structural unemployment we create programs like welfare to ensure a minimum standard of living for every member of society. For many, the system is what they know and they lose motivation to upgrade their skills or change their living situation to try and become active members of the economy. Some even argue that there are disincentives for people in this lower income stratum to work. The best many can hope for is minimum wage. But along with welfare payments may come other benefits, such as rent subsidies, childcare expenses, food supports, and other programs that are not available if you are working. So if you cannot make beyond the minimum wage, there is often a disincentive to work.

So what happens in a depression if the government goes bankrupt? What happens if it simply can't send checks to people on welfare? What happens if old age security payments stop? What if those areas of cities where many people on social assistance live see many people unable to pay their rent and their heating and electricity bills? What if people can't get enough food to eat? This is not a group of people who have been taught independence. They are unlikely to react well to the loss of their basic economic status. Will these areas remain socially stable, or will things get rough?

Remember, governments can go broke. If people lose confidence in the government's paper money or its ability to repay its debt through bonds, it may quickly find itself out of money. Programs will have to be trimmed very fast, and taxes will have to be raised. The stability of many segments of our society will be fragile.

In the 1930s the rural/urban mix was very different than it is today. In the 1920s many rural dwellers began moving to cities to take jobs manufacturing goods that were becoming more popular, like cars and appliances. The rural makeup was agricultural, with a large number of small farms. The mega-farms and industrial agriculture of today didn't exist then.

So when the stock market crashed in 1929 and the economy slowed and entered the depression in the 1930s, there were still many small farms for some urban dwellers to go to work on. Sons and daughters had left the farm to seek their fortune and earn a higher wage in the city. When the jobs disappeared there were still farms that could reabsorb them. Farm life at the time was still very physical. Much human muscle went into the work. So farmers were happy to have more strong backs at the farm to pitch in on the work.

The job mix was different in the depression of the 1930s. Many Americans still worked in manufacturing, in jobs where they made things. These goods were not only consumed in North America but were also exported around the world. Today 70% of U.S. GDP is consumption. The U.S. economy is built on buying "stuff," much of it made somewhere else. America has traded away a large percentage of its manufacturing sector to developing economies and the U.S. workforce has changed. There are many American workers employed in the knowledge and service industry, but many have lost high-paying and often unionized manufacturing jobs and moved to lower-paying service jobs and work in the retail industry.

I believe that in the future many levels of government may become increasingly paralyzed in their attempts to respond to the multiple and cascading challenges they will face. They will have increased demands made on them at the same time as they have reduced resources to help. It will be the loudest and most powerful that will get the first response from government. So it's time to stop assuming that you have a friend in government who will help you out. You don't. You're on your own and have to behave accordingly.

To illustrate the point, let's pretend you're Arnold Schwarzenegger, the Governor of California. You govern a great state with a lot of people committed to doing something about climate change. You want to make them happy by passing tough laws to inspire people to move towards having a smaller carbon footprint, even though most of the state is sprawling suburban development that is very car-dependent and very far from sustainable.

Problem is, right now much of that housing is collapsing in price and many people are having a tough time paying their taxes and keeping their homes. Across the country house prices are in a freefall. Southern California home prices fell a record 34% in November 2008 to a median $US285,000, while home prices in the San Francisco Bay Area in the United States plunged a record 44% in November, 2008 and the median price fell to the lowest since September 2000.[1]

Many people are "underwater" on their mortgages, meaning that their mortgage is worth more than their house. The *San Francisco Chronicle* says: "Two reports suggest that state and federal public assistance programs may be overwhelmed by the growing ranks of families seeking help during the recession."[2] So it sounds as if people are going to be relying on the State of California to help them out. As unemployment grows from the failing U.S. economy, states like California will soon exhaust their unemployment insurance funds and be forced to borrow from the federal government to continue paying benefits. This is really going to put pressure on the budget, just as revenues are falling.

So the state has to borrow money to pay for these expenses. But the interest it has to pay on those bonds has skyrocketed, because investors think it's possible the state might go bankrupt, so they're demanding a premium on the money they lend to it.

That's not surprising, since as the *LA Times* reports that the state's budget gap could reach $42 billion by 2010. That's nearly half its $86 billion in tax revenue. The state has had to go to the federal government to ask for money to help it get through this crisis, but the Fed has been lending out hundreds of billions of dollars and sooner or later they're going to have to stop or default on their loans.

And this isn't the only challenge facing the governor. Like all North American jurisdictions, California has an aging infrastructure that needs billions of dollars of repair. In an earthquake-prone part of the world, you better believe it's an urgent need. We're all getting used to those images of the wildfires that now seem to rage in California during fire season, which now lasts all year. Fighting fires requires lots of water, but as you discover on the website of California Water Crisis (www.calwatercrisis.org) the water supply is also in rough shape. Water supply can no longer meet the state's demand. Last summer the governor declared water emergencies in many parts of the state. This lack of water not only hurts the economy, it affects the state's ability to grow food and get water to its cities. Climate change is exacerbating this problem.

Let's not forget that the problems of peak oil and peak natural gas haven't gone away. California uses massive amounts of oil to move its population around on freeways and uses massive amounts of natural gas to power its energy grid. Both of these energy sources have hit their peak and are in decline. This is going to have a profound impact on the lives of every North American.

Here's how *TIME* magazine summed up California recently:

As 2009 settles in, California isn't quite the Golden State anymore. School districts are expected to lose billions of dollars in financing for improvements and development, and health-care services for the elderly, infirm and poor will most likely deteriorate. State employees are facing payroll cuts, unpaid leaves and a hiring freeze. Money for firefighting in parched Southern California is drying up, as is financing for levees in flood-plagued northern environs of the state. And that's just for starters as California faces a budget deficit of more than $41 billion over the next 18 months.[3]

So you've got an unprecedented number of crises all converging at the same time—environmental, water, unemployment, energy, housing, economic—and your state is basically bankrupt. You don't have the money to rectify the problems and to borrow it is going to cost you more than ever.

California has 36 million residents, is the 8th largest economy in the world, and has a whole heap of problems. Suddenly being Arnold Schwarzenegger doesn't seem so great. What would you do? Where would you start? Who would you help first?

As is happening to governments all over the world, challenges in California are converging all at once. Dealing with them is going to be enormously complex. From the standpoint of the average citizen, the message is clear. You will not be able to rely on governments the way you have in the past. There are just too many demands.

You need to accept this and come up with a "**Plan B.**"

For years I have been giving my Thriving During Challenging Times workshops and have divided them up into "soft landing" and "hard landing" scenarios. In my soft landing scenario I suggest that although we do have many challenges hopefully governments, businesses, and individuals will be able to make the changes necessary to deal with these multiple converging problems. My soft landing recommendations are therefore fairly non-radical and involve things like looking long term at how your job will be affected by peak oil and evaluating where you live in relation to increasing energy costs.

The "hard landing" section of the workshop has a more urgent plan of action and it's the one people don't like to think about too much. We have lived for a long time in a relatively safe and stable environment. Wars are fought somewhere else. Civil dislocation is something that happens in other countries, not ours. So when I suggest that if governments can't deal with all the challenges, and if businesses don't adapt fast enough, we're all going to be dealing with a radically different reality. My recommended course of action is therefore much more aggressive and urgent.

Several other countries in recent history have already experienced the challenges we're about to have to face in North America. As Russia's experiment in communism was coming to an end in the late '80s, it was plunged into a period of huge social and economic upheaval. The state-run economy was in shambles and it took many years for a more capitalist model to take hold. Russians saw a huge drop in their standard of living, to the point where life expectancy went down substantially.

So Russia has already experienced some of the upheaval that North Americans are beginning to encounter. Dimitry Orlov has written extensively about how Russians dealt with the jarring change in a much better way than I think many North Americans will, and much of this is related to expectations. Russians were used to living without much and were accustomed to shortages of things like food and fuel. These reduced expectations were a great defense mechanism in dealing with the changes they experienced.

There were several characteristics of the Russian system that helped them out. First, most Russians did not own homes. They lived in centrally planned, government-owned apartments. Luckily Russian authorities were able to keep the heat and lights on, which meant that people had refuge from the storm. Russians also had a very low percentage of personal car ownership because they did not have the income to own one. And again, the Russian authorities were able to keep operating the public transportation systems which the bulk of the population depended on. So Russians were able to stay warm and get around, which helped them through the difficult transition to capitalism.

One of the other important characteristics of reduced expectations that many Russians had developed under Communism was that they were accustomed to food shortages. Most of us have seen the photographs of bare store shelves with no bread or meat during the days of Communism. Many Russians already had a strategy to deal with food shortages and had a small plot of land in the country where they could grow some of their own food. This was a crucial part of surviving the wrenching transition to capitalism that many Russians experienced.[4]

The Russian economic collapse impacted many of the communist countries Russia had long supported, most notably Cuba, which imported all of its oil from Russia because of trade embargos by the United States. When Russia collapsed, Cuban oil shipments stopped almost overnight. Cuba, in other words, has already experienced "peak oil," and in a much more abrupt way than we will. We have entered a phase of gradual decline in world oil production which will see less and less oil available each year. But it won't stop dead as it did for Cuba.

Cuba undertook a number of radical steps to deal with this. The first was a radical transition to organic growing, without petroleum-based inputs such as insecticides, pesticides, and herbicides. Cubans also had to reduce the scale of their agriculture, since there was no longer diesel fuel for tractors. They replaced fossil fuels with physical labor. Since they were using more labor, a larger percentage of the population was now devoted to agriculture. It simply had to be, and it illustrates my previous point that unlocking the fossil fuel genie freed up humanity from doing the "grunt work" to engage in other pursuits.

Cuba also encouraged the development of urban gardens. Since Cubans didn't have fuel to transport things, it became important to grow the food closer to the end users. Any vacant lot was turned into a garden, and people were encouraged to start their own gardens. Cubans also resurrected old skills, the ones they had relied on before diesel and chemicals replaced them. This included an emphasis on crop rotation and reintegrating animals into the agricultural cycle. Suddenly farmers, particularly older farmers who had this knowledge to share, were held in high esteem. Isn't this a great thing? Here in Canada I often see signs in rural areas with slogans such as "Did you eat today? Thank a farmer." These signs compensate for the lack of respect so many people in developed nations have for the most essential profession of all, that of growing food. We all eat but rarely take the time to appreciate the effort of the farmer who grew that food.

Other parts of Cuban life changed as well. The standard of living went down, which again emphasizes the importance of oil in the wealth of nations. Cubans rode bicycles and took public transit in even larger numbers and personal vehicles became a luxury few had access to. People lowered their expectations. This is a common theme when you talk about peak oil and its impact on western economies. It's a scary topic to bring up, but the impact of less oil on our lives is going to be a lower standard of living. You can appreciate that a politician who was honest about the fate of Western democracies and ran on a policy of the necessity of the electorate to "reduce their expectations" would be unlikely to get voted into office. But this is the reality of the world we will soon inhabit.

If you have an opportunity to read or watch anything about Cuba, I would strongly encourage you to do so. *The Power of Community: How Cuba Survived Peak Oil* (www.powerofcommunity.org) is a film which documents Cuba's challenges connected to the collapse of the Soviet Union and is an excellent example of how people can pull together to overcome adversity. Cuba still has a lower infant mortality rate than the U.S., a higher literacy rate, and trains the greatest number of doctors in the world. If you have a chance to watch *Sicko*, Michael Moore's exposé of the U.S. health care system and the people who fall through its cracks, you will be interested to see the Cuban health care system welcoming Americans who were first responders to the 9/11 attacks and now suffer huge health effects from the asbestos and other toxins they inhaled. These Americans live in the richest country on the planet and yet have no health care to help them. Cuba, which is one of the poorest nations on the planet, welcomes them happily. In movies like *The Buena Vista Social Club*, Ry Cooder's film about bringing Cuban musicians to America, you notice time and time again that Cubans seem to be a very happy people. They don't have a lot, they do not have nice houses and cars, and yet they seem content.

This is the theme of this book. It's not about what "stuff" you have, it's all about you doing things that have meaning and bring you joy. The challenges I've outlined in Part II should change your notion of what's normal and what your expectations should be. But if you can change those expectations and find new ways to define yourself and your happiness, then all this bad news is a huge blessing.

So let's get onto the fun stuff and start your **"Plan B."**

PART III
Solutions

I wanted a perfect ending, now I've learned, the hard way, that some poems don't rhyme, and some stories don't have a clear beginning, middle and end. Life is about not knowing, having to change, taking the moment and making the best of it, without knowing what's going to happen next. Delicious ambiguity.
Gilda Radner

If you don't like something, change it. If you can't change it, change your attitude.
Maya Angelou

The future isn't going to look like the past.
We learned earlier that the challenges facing the planet seem to be piling up. The U.S. economy shed more than half a million jobs in November 2008 and continued to lose between 600,000 and 700,000 jobs per month in the first quarter of 2009. More than 30 million Americans are on food stamps while one in six are behind on their mortgage payments.

The American malaise is spreading around the planet, turning a prolonged economic expansion into a dramatic contraction, which some economists are suggesting is comparable to the Great Depression. Oil and other commodities, which enjoyed historic highs last spring, have plummeted.

The global climate change problem hasn't gone away either. In fact as energy costs were rising we began to see a switch back to more carbon-intensive fuels like coal. Climate scientists are warning we are at or very close to the tipping point. They're starting to discuss "geo-engineering," where we actually have to try and take remedial measures to clean up the mess in the atmosphere.

What is very clear according to people like Thomas Homer-Dixon of the Centre for International Governance Innovation in Canada who examine the big picture is that the future isn't going to look like the past.

Which means it's time for you to stop thinking it is. That retirement to the beach or golf course with ample funds and plentiful and cheap energy to make your life easy may not be going to happen. In fact, for many who have watched their retirement funds drop precipitously in the financial meltdown, this reality is already hitting home. Some might argue that the value of those retirement plans was artificially inflated during the last 20 years without the underlying economic strength to sustain it. If that's the case, it was easy come, and now it's easy go. That's not an acceptable or particularly comforting explanation to most people, but it may be the new reality.

We seem to be moving into a period of dramatic shocks to the systems that we've developed as humans. Shocks to our economic system, shocks to the climate we live in, and shocks to the energy systems that provide North Americans with such a high standard of living. All the systems we've developed have become increasingly tightly linked, from globalized trade to computer networks to just-in-time systems of food delivery. According to Homer-Dixon, systems that are so closely linked are not very resilient to shocks. A shock to one part of the system can have unexpected consequences at various other parts of the chain. In April of 2009 someone cut four fiber optic cables in the San Francisco area, leaving tens of thousands of customers without Internet, phone, or cell phone service. This level of interconnectivity leaves the whole system more vulnerable to a system-wide crash.

So I would suggest it's time you started to make yourself more resilient to shocks.

The following chapters are a road map to guide you on your way. They look at all the major "needs" that most humans have. Many of these are located on the lower rungs of Maslow's hierarchy of needs: the need to eat, the need to stay warm, and the need to stay safe. The future will require many of us to spend much more time and energy dealing with these issues. But this is not to say that you won't also be involved with the higher rungs at the same time. In fact, you may find that dealing with basic human needs brings a new joy and sense of fulfillment to your life and that those higher-level needs start looking after themselves.

Spending less time commuting to a distant job and working hor-

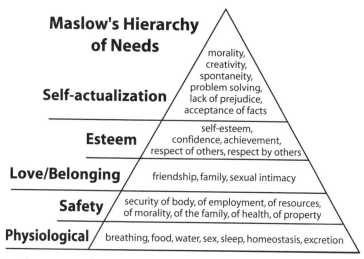

Maslow's Hierarchy of Needs

Self-actualization — morality, creativity, spontaneity, problem solving, lack of prejudice, acceptance of facts

Esteem — self-esteem, confidence, achievement, respect of others, respect by others

Love/Belonging — friendship, family, sexual intimacy

Safety — security of body, of employment, of resources, of morality, of the family, of health, of property

Physiological — breathing, food, water, sex, sleep, homeostasis, excretion

rendous hours may have been affecting your ability to enjoy as positive a relationship with friends and family as you would like. With less income you'll have more time to spend growing your own food and enjoying other people's company, both of which are free.

While many of us have defined ourselves through our jobs, we have often felt underappreciated and ignored in the corporate world. Superiors haven't respected us or listened to our ideas, which can affect our self-esteem. Producing your own electricity with renewable energy and growing your own food is incredibly invigorating. There is nothing like the feeling of accomplishment you get from taking that first basket of vegetables out of the garden or watching your solar monitor tell you how much electricity your solar panels are producing.

And for many, the business world has caused a great moral dilemma. We often know the company we work for conducts itself in ways that we don't agree with. Perhaps its carbon footprint is way too high. Maybe it sources its products from developing countries that don't have the same environmental and labor standards that we have here. But you rationalize that you need an income, so you have to grin and bear it. Sometimes having to leave a job like that, whether voluntary or not, represents a huge opportunity to start doing something that you truly believe in. Something that makes the world a better place and helps other people, or at least doesn't do any harm.

This can be incredibly freeing, invigorating,… and good for the soul.

In fact every suggestion that follows is a good thing on a number of levels.

First, these ideas will help you and your family deal with the increasing shocks to the system that are going to accelerate in the future. The suggestions that follow will make you less vulnerable to energy, food, and financial shocks. They will make you more resilient to these crises.

Second, they will be good for your health, both mental and physical. Riding your bike more often is good for you. Growing your own vegetables is excellent cardiovascular exercise and, as so many people have discovered with flower gardens, it is an incredibly satisfying and enjoyable activity. Hoeing weeds lets you "zone out" and reflect on things that make you happy.

And finally, everything that follows is good for the planet. It will help you reduce your footprint on the planet, and the planet desperately needs you to do this, now! Using renewable energy to make yourself more energy independent also helps you reduce how much coal is burned to make the electricity you use in your home. Eating food grown in your garden or by a local farmer dramatically reduces the energy required to get food to your plate and therefore lowers the carbon produced. Harvesting your rainwater rather than having it run down your driveway into a city sewer reduces the stress on your municipal infrastructure and helps to eliminate raw sewage being washed into local watercourses when the rainwater sewers become overwhelmed during storms.

I would suggest you start developing a "Five-Year Plan." Take the information provided here and chart a course for your next five years. Have a goal, whether it's to buy a smaller house closer to work so you can take transit or to buy a place in the country so that you can grow all your own food. Figure out what it's going to take to get there, and then develop a plan. Come up with realistic financial goals and timelines, and then put them on paper. Keep referring to the plan. Put it on the fridge. Keep checking it. Put it on the breakfast table next time you're considering a big purchase. Do you really need a new couch or TV, or will the existing one make it through another year? If it will, put that $1,000 right into the "Five-Year Plan Account." You won't be so disappointed about watching movies with those black bands at the top and bottom of your "non-widescreen" TV anymore. Those black bands represent that house in the woods you've always fantasized about. That $1,000 you didn't spend just brought you that much closer to the goal.

There are numerous reasons to embark on a program to make yourself more energy, food, and financially independent. This book will point them out as we go but will keep coming back to the main benefits: it's good for your health, good for the health of the planet, and good for your resilience to a less certain future.

9　Where to Live

Four walls, a roof, a door, some windows
Just a place to run when my working day is through
Dixie Chicks, A Home

Where you like to live is a very personal thing. Lots of things affect it. Where you work. If you want to be close to family. If you like the convenience of living in a city, or if you long for the wide open spaces of the country. Some people like to be around people and freak out when they find themselves isolated. Others are almost allergic to the crowds. John Cougar Mellencamp "can breathe in a small town," but Frank Sinatra sang "If I can make it there, I'll make it anywhere, it's up to you, New York, New York." It takes all kinds.

Most North Americans choose to live in and around cities. In 1950 about 60% of North Americans elected to live in urban centers. By 2005 that had grown to over 80%. Over the past century more and more people have chosen to move from the country to urban centers. They've been drawn by jobs, cultural pursuits, and, of course, $5 lattes at Starbucks.

Cities can be very sustainable places to live. Lots of people hear of someone living off the electricity grid in the country and make the assumption that this is the definition of sustainability. If they live in a place where lots of the electricity is generated from coal, there's no doubt that the carbon footprint of that off-grid dweller is going to be much smaller when it comes to electricity. But how do those people earn their living? Are they retired, or do they own a home-based business? If they don't have to leave their rural home very often, then they're doing the right thing for the planet.

What if they work in town though? Suddenly their half-hour or one-hour commute into an urban area has negated the benefit of their solar panels. And what do they use for their sources of heat? If they have simply shifted all of their heavy electricity loads, namely heat sources, to propane, then they really haven't changed their energy footprint significantly. In fact, someone living in a city and using electricity from a nuclear plant for their cooking and hot water is probably further ahead in terms of the CO_2 that they're putting into the atmosphere.

What if that urban dweller also uses public transit to get to work rather than driving as so many rural folk do. And what if they live in a condo or apartment rather than a detached home. Think of the heat loss in an apartment. Generally, you only have the outside wall exposed to the elements. Three of your four walls abut other inside walls, so you don't have the heat loss you would have in a detached house where every wall is exposed to cold outside air.

I would suggest someone living in an apartment in a city, taking transit to work, using energy as efficiently as they can, and shopping at the local farmers' market has a much smaller footprint on the planet than many off-gridders. I'm not saying one is better or worse, just that there are many shades of gray in this debate. The key is to determine how you can reduce your impact.

When it comes to dealing with the challenging times we're confronting, where you live is going to have a major impact on how you deal with the challenges. Neither is better or worse; each has advantages and disadvantages; so let's look at the two main options.

The two best choices are living in the city or living in the country. The wrong choice is living in the suburbs. They are less and less desirable as many of the challenges converge on us. Suburbs were built using a model of urban planning that is premised on large amounts of cheap energy. The homes are far apart, and they are located far from shopping and jobs. You have to drive everywhere. Since the homes are so spread out, it is difficult for local governments to service them properly with transit. Transit works best when people live densely, as in a city. The suburbs are so spread out and thinly populated that it is difficult to provide good transit services. That leaves you reliant on a car, and with peak oil driving the cost of fuel up from now on, this is going to get very expensive.

Suburbs tend to have single, detached, large, and hard-to-heat homes, making them the least desirable place to live in the future. The infrastructure in the suburbs will be an issue. For every mile of sewer or water line, far fewer residents are serviced in the suburbs, so a city government with diminishing revenue is apt to put money where it will do the most good, and that's in the more densely populated city areas, not the suburbs. If you live in the suburbs now, perhaps you should be planning a move in your Five-Year Plan, either to the city or to the country.

Living in the City
Pros

The great advantage of living in the city is that many of us already live there. It means less disruption and more continuity. If we've got lots to deal with in terms of the economy and our jobs and food and energy issues, it's better that we don't also have to deal with the jarring impact of a move to a radically different way of life. People accustomed to having a myriad of shopping and services close at hand often have trouble adjusting to life in the country where that level of service just doesn't exist.

With most citizens living in cities, governments at all levels will have a natural tendency to devote resources to those areas. Governments will be pulled in many directions as multiple challenges converge on their resources, but they're going to have to make sure they keep the majority of their citizens happy, especially as long as North American governments are democratically elected. It only makes sense to keep the majority happy if you want to get re-elected.

The other advantage of living in the city is its proximity to jobs, or potential jobs. Businesses tend to cluster in urban areas to be close to a work force and services such as electricity, water, sewers, and high-speed Internet.

So if you have a job, it's likely already in an urban area. If you are looking for work, the greatest number of potential employers is in the city. If your job is accessible by transit, this is even better.

Transit is going to be one of the priorities for governments, In fact, as the world starts to become aware of the implications of declining oil production, governments will finally start to make the investments in transit that they should have been making all along. In the spring of 2008, as the price of oil was approaching $150 a barrel, transit ridership throughout North America increased dramatically. If the price of oil had stayed there much longer, transit authorities would have had to begin a crash program of infrastructure investment to add the trains, buses, and subway cars required to move the higher number of riders.

With the price of oil dropping, we have had a temporary reprieve from the shock that oil at $150/barrel caused. But this is fleeting; it will only last so long. Soon the market will return to its recognition of just how precious the remaining oil in the ground is, and the days of happy motoring will be over. Transit will become the most cost effective option for a large part of the population, and you'll find the best transit in the cities.

As Russia went through its jarring transition from communism to capitalism, one of the things that allowed life to go on as usual was the fact that most Russians didn't own a car. They were used to public transit as their major mode of personal transportation, and even with the problems that befell the transitional governments in Russia they were able to keep the buses and trains moving, which kept people working and the system functioning.

You should expect there to be jarring impacts in the North American car-based transportation model, but I believe you will be able to count on the government to keep public transportation moving. This means you'll be able to get to your job and to go and buy food.

Food is going to become an increasingly bigger part of your future. It's going to get more expensive and it's going to be harder to come by. As discussed earlier, it's going to be harder to grow as much food without abundant natural gas for fertilizer, and because there are so many fossil fuel inputs into our food supply, the rising cost of energy is going to hit the price of food in a variety of ways. In the city, you are pretty much completely dependent on someone else to provide your food. You can certainly have a garden if you own your own home, and you can practise intense gardening to try and squeeze every ounce of food out of the soil, but the reality is that in most urban areas you simply won't have enough land to make a huge dent in your food budget. You are still going to need to buy food from someone else.

If our current food infrastructure remains relatively intact, stores will still have a variety of food, but the variety may be greatly reduced. You may you have to change your approach to eating and substitute food choices you wouldn't have considered in the past. Heavily processed foods will likely be proportionately more expensive, because all that processing requires energy; so you'll likely need to start eating lower on the food chain, meaning less reconstituted potato chips in a cardboard tube and more potatoes. And the most economical source of those potatoes in a carbon-constrained future will be local farmers.

Local farmers are going to want to maximize the return on their investment of time and money, especially their fuel costs. What they'll probably realize is that it makes much more sense for them to market their food in areas with the highest population density, and that will be cities. Farmers' markets are becoming common fixtures in urban areas and they are the beginning of a model of food distribution that is going to become the norm. Farmers are going to drive to an area of high density, and people in that city will have to find their way to that market, be it on foot, bicycle, or transit. This form of transportation will reduce how much can be carried, so people will be eating fresher food that they purchase more often.

So living in a city and using a farmers' market is going to be better for you by improving the quality of the food you eat; it's going to be better for the planet, reducing the miles that food travels before it gets to your plate; and it's going to help you minimize the effect of rapidly rising energy costs.

It is surprising to learn that people who live in cities are much healthier than people who live in the country or the suburbs. People in the outlying areas too often substitute liquid hydro-carbons (gas, diesel) for human calories. They drive more for work and shopping, and even though they have the opportunity to get exercise by cutting large lawns or cutting firewood they tend to use ride-on lawnmowers rather than push mowers, and miss the opportunity for personal exertion. City people know the hassle of driving. Traffic makes moving around a city slow, and parking adds to the expense and inconvenience of a trip. Taking transit often makes more sense, but the bus or subway stop isn't usually right outside the door, so you tend to walk more. If your favorite coffee shop is four blocks away, it doesn't make sense to drive or take transit, so you put on some comfortable shoes and get walking.

There are lots of other advantages to city life. You'll have more access to cultural events, especially if the economic downturn reduces advertising revenue and we see our 500-channel cable TV world start to contract. You can count on there being lots of creative people in cities who are happy to use entertainment as a source of income. You'll be closer to stores that will help you save money, like second-hand stores and pawnshops, and you'll be able to find a library nearby with an almost infinite amount of free entertainment available in the pages of its books. If you're reading this book on loan from a library you already understand this concept. If you purchased our book, we thank you.

Cities will always be vibrant places to live, with a variety of creative people teaming up to use their talents and resources to deal with the coming challenges.

Pros of City Living - Summary
• Public transportation infrastructure a priority
• Ability to walk and cycle for work and food
• More economic incentives for farmers to bring food to areas with denser populations
• Access to cultural activities

Cons

Living in the city will have some potential downsides. One of those is food. While I think it will be a good place to live if farmers can get their produce into a central market area, what if they can't? What if the effects of peak oil are more dramatic than we anticipate and farmers simply don't have the diesel to put in their trucks to bring their food to the city? If we get to the point where farmers have to go back to relying on horses, this is going to pose a problem.

In the old days, a farmer could hook up his wagon and set out for town with his food to sell or trade. The problem now is that suburbs have made the area around cities a wasteland of strip malls and super highways. Lots of good farmland got chewed up in this process, but it also created a fairly cumbersome barrier for someone from a rural area trying to get to the downtown area without the use of a fossil-fuel-powered vehicle. A horse and wagon just couldn't realistically do it. They certainly can't get there and back in one day and we don't have facilities in cities for horses anymore. So if we reach the point where farmers can't get food downtown to the farmers' market, you'd better have a Plan B that includes you living closer to the farmer.

In the city you are very dependent on other people, not only for your food but also your electricity, your heat, your water—everything that makes your life convenient. When we started our migration to urban areas, we began giving up our independence and trading our labor for an income based wage that gave us enough money for life's necessities, like food and water and heat. This arrangement has worked for decades, and cheap and abundant energy has made it work well. Power utilities can build power stations and source fuel (coal, natural gas, uranium) that allows them to produce electricity cheaply enough that it's a fairly small part of most people's paychecks.

Other utilities have been able to purchase natural gas and pump it through pipelines to heat our homes and apartments and businesses. As long as it is cheap they can make a profit and we don't chew up our whole income staying warm. The city can purchase electricity at such a low cost, along with the chemicals needed to process your water and supply it to you at such a low rate, that most people hardly notice things like water and sewer services in their monthly budget.

This model of trading your labor for income to purchase these items works as long as products and services are inexpensive and as long as energy prices stay low enough that there's room for profit. The model goes out the window as energy prices rise and the costs of building and maintaining the infrastructure to keep you supplied with these essentials gets too high. This is going to be the model for the future. A declining amount of income for many North Americans will occur at the same time as the costs of many basics rise dramatically.

People in the country have some of the same problems, but they are more likely to have control of the necessities that city dwellers have farmed out to someone else. Country people generally have to provide their own water from a well. They know how to get the water out and since it usually depends on electricity many already have a backup system like a generator to keep it flowing even if the local utility company stops providing power. In a city you are completely dependent on essentials that you have no control over in either their delivery or their price.

As governments try to adjust to this new environment of rising energy costs and declining tax revenues as economic activity declines, they are going to have to reduce some or many of the services they currently provide. Many of the people who receive this assistance live in cities: people who are unemployed, on welfare, or on disability pensions. As governments have to make choices about how to spend money, this group may find themselves receiving less support from the government just as the cost of living is rising because of energy prices. Some of these people are not going to be happy about it. In the last 20 years we have seen a huge increase in the gap between rich and poor Americans.

In fact, the Organisation for Economic Cooperation and Development (OECD) reports that:

> The United States is the country with the highest inequality level and poverty rate across OECD, Mexico and Turkey excepted. Since 2000, income inequality has increased rapidly, continuing a long-term trend that goes back to the 1970s.... The distribution of earnings widened by 20% since the mid-1980s which is more than in most other OECD countries. This is the main reason for widening inequality in America.[1]

In an article entitled "Americans See Widening Rich-Poor Income Gap as Cause for Alarm," *Bloomberg News* reported:

> The portion of national income earned by the top 20 percent of households grew to 50.4 percent last year, up from 45.6 percent 20 years ago; the bottom 60 percent of U.S. households received 26.6 percent, down from 29.9 percent in 1985, according to the Census Bureau. Meanwhile, average pay for corporate chief executive officers rose to 369 times that of the average worker last year, according to finance professor Kevin Murphy of the University of Southern California; that compares with 131 times in 1993 and 36 times in 1976.[2]

As more and more people join the ranks of the poor and the gap between them and the rich increases, and as the government cuts back on social programs, they're going to be increasingly frustrated. This will be manifested in the form of increased crime and civil unrest. The poor will look for someone to blame and it will either be governments or authority figures or those who seem to still be doing well. This is basic human behavior. People want to see some degree of fairness. The last 100-year bonanza of cheap energy and abundant economic activity to support government is about to end, and for many it will be a hard pill to swallow.

Some will act out of anger and some out of desperation. We can all appreciate how a person might feel and behave if they or their children are hungry or cold. If they can't get help from the government, they will search for a solution somewhere and resolve it, potentially through illegal means.

This just means that cities are going to become more dangerous places to be. Over the last 20 years we've seen quite dramatic decreases in the crime rate in many major North American cities. I believe the days of this trend are over. There are going to be increasingly angry and desperate people committing crimes, and police departments are going to be stretched to their limits and unable to stop much of it. Ideally their budgets should increase to deal with this, but they'll in fact be going in the opposite direction as cities grapple with less money.

While many "doomers" would suggest that urban centers will soon resemble a *Mad Max* movie, I believe we'll have the resources to keep them quite livable, but living there will bring us into much closer proximity to people at their wits' end and willing to do whatever it takes to get by.

> ## Cons of City Living - Summary
> • Food harder to come by
> • Increased dependence on someone else for necessities such as water, electricity, food, and heat
> • Social unrest more likely as income gaps widen and economies have less excess to help those less privileged

Living in the Country
Pros

For many, the first response to challenging times is to want to "head for the hills." Here are some of the reasons that may not necessarily be such a bad idea.

One of the major downsides to urban living is going to be a huge advantage of country living, and that's food and your access to it. Chapter 13 is devoted to the subject of food, so I won't dwell on it now. The executive summary overview of it is simply that food is going to require a much greater effort on your part and it's going to be easier to grow or access food in rural areas. Most rural homes have more property and hence more room to grow your own food. You also tend to have access to some of the important elements required for healthy gardening, such as manure. Commercial fossil-fuel-based fertilizers are going to be expensive and hard to come by, so finding natural fertilizer is important. As energy becomes more expensive, finding it locally is going to be even more crucial. And if you can't find it on a farm nearby, you might choose to keep animals and make it yourself.

A well-managed small-scale farm that includes livestock like horses, cows, pigs, and chickens forms an excellent basis for a self-sufficient rural oasis. You can feed the animals on grasses and hay on land that may not be prime for growing vegetables and then use their manure to enhance your vegetable garden and increase its production. Most of us have no background in growing food because we were born and raised in a time when cheap energy allowed farmers to grow far more food than they needed and sell the surplus to us. For us, food comes from grocery stores and restaurants.

Well, that simply isn't the case. Food comes from the ground. It has to be planted and nurtured and weeded and watered and harvested and stored and canned and frozen and it requires an enormous amount of time and effort. It's incredibly rewarding, but for most of us it's a foreign concept that we're about to become intimately in tune with, very soon.

If we do live in the country and find a way to earn an income and choose not to grow our own food, we will be in close proximity to people who have food to trade, namely farmers. Farmers have the skills to grow a surplus of food and so will be willing and able to sell or trade some of that extra to their neighbors. It is much easier to trade with a neighbor than to have to transport it to an urban area farther away. The price may not necessarily be significantly less than you'd pay in the city, because the farmer knows that people in urban areas may be able to and have to pay more for essentials like food. But the odds are that some of your neighbors will have food to trade.

Another advantage of living in the country comes from the trend we see now, which is that rural residents are more likely to adopt renewable energy than urban dwellers. There are lots of reasons for this. People in the country like to be more independent and solar and wind power allows this. Electricity can be more problematic and power interruptions more common in the country, so having a system to keep your appliances working during an outage is top of mind with many country people. This happens out of necessity for most. To pump water, you need electricity. So when the power goes out, the water stops flowing. A human can go only 72 hours without water, and for most of us, having a toilet that flushes is pretty important.

People in the country are often more familiar with flooding than city people. Without storm sewers, rural basements can have a tendency to flood after big rainstorms or during the snowmelt in the spring. The way to deal with this is to have a "sump well" where the water collects and a "sump pump" to pump that water out of the basement. The sump pump will cycle on and off as the water rises, drawing the water down and shutting itself off. This pumping can last days or weeks until the water stops flowing into the sump well. As long as the electricity keeps flowing, the water keeps being pumped out. But as soon as the power goes out, the water starts backing up and the basement gets flooded. Rural people who have had to replace carpet or drywall once usually vow that it won't happen again and they make sure they have a backup system to keep that pump on. This can be a backup generator or a renewable energy system with battery backup.

Having more room on their property seems to inspire more people in the country to want to put up solar panels or a wind turbine. While you don't need to put solar panels on a tracker, for many the ability to track the sun as it moves across the sky adds to the attraction of solar power, and a tracker needs a wide swath of unencumbered horizon. There's no reason for many urban dwellers not to put solar panels on their roof, but the greatest number of early adopters have been rural dwellers. Wind turbines require a large unrestricted flow of air to be most effective, so the turbulence encountered in urban areas make them a bad option for wind power in the city. The more open unencumbered areas often found around farmers' fields and grasslands make wind much more attractive, and rural homeowners often make wind power part of their energy systems. So living in the country seems to inspire and be conducive to homeowners integrating renewable energy into their living arrangements.

I cannot emphasize enough the intimate relationship most rural residents have with water. No matter where you live you need to drink water, and it's very nice to have water to flush the toilet, wash clothes and dishes, and keep yourself clean; but city folk often take it very much for granted. Someone else provides it and the assumption is that it's limitless and safe. Having a well for your water makes you acutely aware of the finite nature of water and the difficulty of ensuring that it's safe. Wells replenish themselves only so quickly, so you have to be aware of how much water is available to you. And with potential agricultural runoff or industrial pollution upstream somewhere in your watershed, you need to have your water tested to ensure it's safe.

With this awareness of water comes the knowledge of the impact of not having it available, whether through loss of electricity to pump it or drought restricting its flow. Country folk tend to have a "Plan B" for what happens when the power goes off or the water isn't clean enough. While I would encourage urban dwellers to be aware of the precarious nature of their water supply, especially during power disruptions, it is much tougher when someone else controls it.

That increased control over the essential operations of a household also applies to heating. Many rural residents are now discovering ground source heat pumps as an excellent and very environmentally responsible way to heat. While it's an excellent option for a city home, many of the first heat pump users had large properties that made a horizontal loop system possible. This will be discussed in more depth in the heating section.

Heating with firewood is also more common in the country. Some people have enough property and a large enough woodlot that they can supply themselves with their own firewood. Those without their own woodlot can usually find local suppliers. Providing you are using an EPA-certified woodstove and burn your wood properly, heating with wood is an excellent, carbon-neutral way to heat and is much more common for those living closer to the source, that being forests. As natural gas and other forms of heating become increasingly expensive, wood will be an excellent option for many people.

Pros of Country Living – Summary
- Ability to grow your own food
- Proximity to farmers to trade and barter with
- Easier energy independence:
  - wind turbines and solar trackers easier to install
  - water you can pump yourself
  - firewood you can harvest or buy locally

Cons

In all things, nothing is perfect, and country life is no exception. The single biggest downside to living in the country is transportation. People in the country drive more. Whether it's to the store, to work, to the restaurant in town for breakfast out, or to the dump, you tend to get there in a car. In a carbon-constrained future, this is going to become increasingly expensive. In a severely restricted post peak-oil world, it could become more than expensive; it could become unaffordable for many. The allure of country life for a great many people has been premised on mobility that cheap oil has provided. Those days are over, so just how attractive the dream of country life is will depend on a number of factors. Do you have to drive to a job to pay your mortgage? How much food do you plan on growing yourself and how much will you have to purchase at the local store? How social are you? How old are your kids? If kids are in soccer and hockey, which entails once-a-week practices and twice-a-week games, how much driving are you going to have to do?

One of the best options would be to have property just outside of a small town, where you could walk or cycle into town, but this isn't always available. It also may be more expensive than property further from town, because it is a more attractive place to live. So where you end up will depend on your budget. And if your budget limitations see you living further away from the nearest trading location, you may be going there less than you'd like.

Not getting into town very often could be one of the major drawbacks of living in the country. Most of us are acclimatized to interacting with other humans. It's one of the reasons we live in cities. So finding yourself more isolated in the country can be a major drawback for many people. Lots of people try the country thing but end up back in the city to be around people and the other amenities that come with city life.

This may be more of a disadvantage in the future as we move from a money economy in which we trade our labor for dollars which we in turn purchase things with to one in which we increasingly trade our labor or the fruits of it for the fruits of someone else's labor. This is based more on a barter model which, with collapsing currencies or rapid inflation or deflation of paper currencies, could become much more attractive and common. While it may not give you the variety we have in our current economic model, it may give much better value. It will also force you to get to know your neighbors better to increase the pool of potential goods and services to trade for. This will happen in the country and in the city, but there's a good chance in the city you'll have a larger potential trading base. The question is whether there are products you need in the city. It is likely that there will more trade in food and basics in the country, while the city may offer more finished goods or luxury items.

> **Cons of Country Living - Summary**
> • Transportation a potential problem
>   - more driving (or longer distances on your bike or horse)
> • Isolation - fewer people to trade with

Some people have looked at another model of living arrangements and have explored intentional communities. While there are no hard and fast rules for this, the model is often based on a fairly large piece of land and a building or buildings which are jointly owned by a number of people. Community members have their own living space and often share a large common area such as a kitchen. Each person has a particular strength like cooking or gardening which they focus on but is also responsible for contributing to all areas of the community, even if it's to a lesser degree. Some communities have people buy shares to become members; others work on arrangements such as working towards ownership or membership.

The concept of an intentional community is an excellent one. It's based on shared responsibility and using people's strengths to improve the community. It's part of the "two heads are better than one" train of thought, only this one can have multiple inputs. There can be challenges though, many of which the communities try to address right up front. They generally work on a democratic model, so that major decisions are voted on and majority rules. Human nature being what it is, personality conflicts can develop and egos can be bruised. If the community has been set up correctly and everyone knows the rules going in, this should help alleviate the problem. But people are people, so inevitably there are going to be challenges with certain personalities clashing or visions of the community's direction coming into conflict with each other. Today, because so many of these arrangements are voluntary, members often come and go. This may very well be one of the best ways for humans to live in the future, in which case the advantages of someone living there may well outweigh the frustration with how it is governed. And there will probably always be gardens to be hoed by hand or firewood to be split for those that need an outlet for their aggression.

What to Live In

This seems like an easy question. You should live in a house. That's the American Dream and what most of us aspire to. I agree. There is nothing like the security of owning your own home. A roof over your head, four walls, and a warm, dry place out of the elements. The question is whether this is the best time to own a home. Many people reading this will already have had this decision made for them because they were unable to keep up their mortgage payments. Or as the value of their home dropped dramatically it didn't make sense to keep paying for a home that was worth less than the amount they owed on it. I'm hoping these people are happy with the ideas presented here and see it as a real bright spot. Going through what they've gone through is traumatic and gut wrenching.

Home ownership has been a very secure investment for most of our lives. Rather than paying rent, you are building equity with every mortgage payment. Over time the price of your home goes up and for many of us this is the largest asset we own. During the last decade this asset was appreciating in value dramatically. And it appeared pretty secure because housing prices tend to go up over time. Even Ben Bernanke said, as he was taking over for Alan Greenspan as the Chairman of the Federal Reserve: "We've never had a decline in housing prices on a nationwide basis." Well there's always a first time and this is it. And unfortunately it's been brutal. While many pundits had been predicting a housing collapse, it was hard not to get caught up in the frenzy. The result is that millions of people have lost their homes and millions more are underwater or upside down on their homes, actually owing more than the home is worth.

There is lots of blame to go around for this mess. Alan Greenspan for inflating the housing bubble. Unscrupulous mortgage brokers for giving mortgages to people who shouldn't have had them. Banks for selling those mortgages to Wall Street to bundle in CDOs and sell all over the world. The credit rating agencies that weren't giving fair ratings on the potential downside to these practices. But ultimately we have to take some of the blame. Regardless of how attractive the housing market was and how relatively stable it had been for so long, there always was risk involved in what we were doing. Owning a massive home that cost hundreds of thousands of dollars and that we hadn't put a significant down payment on was a risky move. Yes, as long as the market continued to go up it was a good bet, but markets can always go down. You have to factor in some downside risk regardless of how remote it is. But how we got into this mess is sort of irrelevant now. How we deal with it is what's important.

In his book *After the Crash*, financial and real estate analyst Garth Turner discusses the two major factors that should be affecting your housing decision right now: the aging demographic of our population; and whether you think we'll have inflation or deflation in the future.[3]

The baby boom generation, born after World War II and up to 1960, is the largest segment of our population and is now getting to retirement age. This generation has known incredible wealth and prosperity and really was born with a silver spoon in its mouth. Many boomers now own large multi-bedroom homes which they won't need much longer. The kids have moved out and they want to downsize. This coincides with the real estate bust which saw large homes on small lots built in every city and suburb in staggering numbers throughout North America. The number of these homes that are unsold and in foreclosure now grows steadily. So at the same time that you have this massive buildup in inventory you have a generation that is about to start dumping theirs on the market as well. This does not bode well for a rebound in home prices in the future, especially for this type of house. Boomers will be looking to downsize and have more compact homes. And as we all become aware of the impending natural gas and oil shortages, energy efficiency will become a desirable attribute which is less realistic in a massive home with vaulted ceilings and marble entrance halls.

From a macroeconomic perspective there are two possible outcomes of today's financial crisis. One is inflation. Governments are desperate to get the economy rolling again and get people back to work, but in an economic downturn people become very cautious with their money and don't want to spend it. This is the correct response (one you should have all the time) and in fact you see personal savings rates climbing from the negative rate they were at during the height of the housing boom. To avoid this the government can reduce interest rates, making it cheaper to borrow and spend, or it can "increase liquidity," meaning print more money to give to people to spend. Economists and Federal Reserve chairs have seen the threat of increasing the money supply and therefore creating inflation as very negative for so long that they now called this "quantitative easing." This is the new term given to printing money and inflating the money supply. Since the interest rate is essentially zero right now, the only option the government really has is to print money.

The Bush administration mailed out checks to Americans to get them spending again, but many just saved the money or used it to pay off debts for stuff they had already purchased. Obama is taking the same course and going on an unprecedented spending spree trying to get the economy rolling again. And the Fed has told us that through "quantitative easing" it is going to start putting more dollars out there to cycle through the economy. This means the possibility of inflation, which can destroy the value of your dollars and investments because if there are more dollars each dollar is worth relatively less. In the 70s and 80s oil shocks and other economic events created rampant inflation that was very hard on consumers. For the last several decades inflation has been pretty much under control. While the powers that be keep changing the definition of inflation and are therefore better able to manipulate the numbers, officially the inflation rate has been manageable. Many economists actually argue that some inflation is good and the best way out of this recession is some good old inflation.

The other macroeconomic outcome of our current economic downturn could be deflation, which can be a very bad thing. Right now many of us are less likely to buy a new home or a new car because we believe that prices are going down or that better deals may be available soon. So we hold onto our money and wait. While that is a sound strategy for you, if everyone in the economy does the same thing it can become a bad thing. With deflation there is the expectation of falling prices and therefore a reduction in the incentive to spend. As long as everyone is sitting on the sidelines and not spending, the likelihood of the economy coming out of its downturn is unlikely. And if everyone is waiting to buy stuff because we believe that prices will continue to fall, this becomes a self-reinforcing feedback loop and the economy just continues to worsen. Then many of those people who made the choice to hold off making purchasing decisions end up losing their jobs, which further reinforces the cycle. Deflation is a nasty thing to have in an economy and it's very difficult to get rid of. You can curb inflation by raising the interest rate and reducing the money supply. Central bankers and governments do not have the same tools to fix deflation.

This was the nature of the Great Depression in the 1930s, and as people's expectations worsened it made it more and more difficult to see a way out of the mess. Then along came World War II, and massive government spending got lots of people back to work, ending the vicious cycle.

So the question is whether you think that a deflationary period is more likely than an inflationary period right now. If you do, then it is obviously the wrong time to be buying a house. In fact if you don't own a home right now you may be better off than someone who does. Rents right now are more likely to stay low because of the huge volume of homes on the market. Some of the houses that are in distress or that builders have built and now can't sell for their real value will just end up being rented. Every indicator right now points to the price of a home continuing to drop, potentially for quite a long time. So if you are renting, great. Save your money and start building up a nest egg to buy your dream home in the country. You have to save your money now more than ever because when you are ready to get back in the market financial institutions are going to be much tougher than they were in the last decade. They learned their lesson. No-money-down mortgages and all the crazy financial vehicles that were available were the wrong way to own a house. To buy a house you have to have saved some money for a down payment and be buying a home you can afford based on your income. When someone has no equity invested in a home it just makes it all the easier to walk away from it.

This housing crisis represents tremendous opportunity for someone renting right now. They should be able to find a decent place to live for a fairly low amount, and this should give them more opportunity to save up a sizable down payment. If you own a home now, though, you are in a different situation. If we go into a deflationary period and it lasts for years the house you're living in now will lose value. A home should really be for shelter and should represent a small part of our financial assets.

But for many of us it's our primary and largest asset by far. In fact right now with so many people maxed out on credit debt, it is their only asset. If your only asset is dropping in value and its salability over the long haul is questionable, then you have to ask yourself whether it is a good investment.

So what I'm suggesting is that perhaps you should consider selling your home and renting until things settle down. I realize this may seem like a bizarre recommendation, but you have to remove your emotional attachment to your home and take a critical look at it as an asset. Since you have probably seen its value drop as houses around you sell for less you know what's going on. We all hope it will hit bottom soon and stabilize and certainly there are those in the real estate business who try and remain shiny and happy and see a bright spot on the near-term horizon every chance they get. But realistically you have to look at just how bad the economy really is. There is a real chance that we're in for a major deflationary period that could be prolonged. When Japan's real estate bubble burst in the early 90s the Nikkei index plunged and sent Japan into a deflationary spiral from which it is still recovering. The Japanese government did many of the things our governments are trying in an attempt to fix the problem, like recapitalizing banks with $500 billion to get them lending and dropping the interest rate to zero. Yet the Japanese economy limped along and the Nikkei today is trading around 8,000 twenty years after hitting its 1989 high of 39,000.

So there is no guarantee that our economic woes will turn around soon. This makes it questionable for you to continue to pay a mortgage on a steadily declining asset. Let's say you own a $500,000 house with a $450,000 mortgage (10% equity) which you sell, ending up with $50,000. Then let's say you rent for five years and after paying off all your other debt you manage to save another $50,000, so you have $100,000 equity to put into your next home. If deflation continues as it very well could, in five years when you're ready to get back in the market you may be able to buy a comparable home for $250,000. Now your equity is 40% and you only have a $150,000 mortgage. Doesn't that sound more manageable? When you're sure the market has bottomed out then you can also be more assured that the price will remain stable or potentially rise. Owning a depreciating asset with a large liability attached to it is a scary proposition, as we're all learning.

Your first reaction is that this is the wrong time to be selling a house. I agree this is not a prime time to be a seller in the real estate market. The question is when will be a good time? One year? Two years? A decade? If you wait too long and the economy remains in the grip of deflation, you'll be putting your house on the market when all those boomers who want to downsize do as well. It could be a double whammy. They cannot stand the thought of selling now because of the value they have added to their home which they will now lose. Some of them will want to hang on, which gives you an opportunity to sell now before things get even worse.

Yes you will take a hit selling now, but if you can get out now without losing a lot of money then you can get yourself back on track to enter the housing market with a half decent down payment at a time when the cost to get back in will be much more affordable. Hopefully the housing market and the economy will be much more predictable then as well, making such decisions easier. This may seem like a crazy proposition, but if the value of your home is dropping steadily and your mortgage is not and there is the likelihood that you or your spouse may lose your job, then a defensive action like this is one scenario you may want to consider. This way you are being proactive rather than waiting for the next shoe to drop and having the decision made for you by someone else like your mortgage holder.

I cannot predict the future. I am just putting one possible scenario out there. You have to do the research. You have to decide whether you think the government will be able to prime the pump and get some inflation happening to turn the economy around. Or whether you think that millions of consumers will elect to hunker down and wait out the storm and send us deeper into a deflationary depression that is very difficult to get out of. If you think that deflation is more likely then you need to think of a home as shelter right now and get out of the one you currently own if it could drag you down financially in the long run. Take advantage of it being a good time to rent and use the money you'll be saving each month to pay down debt and build up a down payment for when you get back into a much saner market.

Not owning a home if you always have might seems like a scary proposition, but it may actually be less scary than owning a home with a large mortgage while you watch the value drop. Yes, you may take a hit to get out now but it may be much less than if you hang on. And perhaps if you're out of the market for a while when you're ready to get back in you'll have a better idea of where the best location is. By then you may be convinced that it's time to move to a smaller community. You may decide it's time to buy land with trees to heat your home sustainably with wood. If you want to stay in an urban area your expectations in a few years may be more realistic and you may realize that the kids can actually share a bedroom to allow you to have a smaller, more manageable house and mortgage. And maybe in a few years you'll be convinced of the reality of peak oil and peak natural gas and base your purchase on the energy efficiency of the house and how you're going to be able to heat it with the rapidly rising cost of fossil fuels.

These are challenging times, which sometimes require a step back to look at the big picture. The various economic foundations that we have held true for a generation are crumbling. The one-time largest corporation on the planet, General Motors, is bankrupt. The Wall Street investment banks have disappeared and taken tens of thousands of jobs with them. The government is playing a massive role in the economy and running up massive debt. These times are unprecedented. Stepping back and thinking outside the box is what will help you navigate the uncharted waters the world economy is in.

Conclusion

No one has a crystal ball on how the future will unfold. In the reference section of this book we've provided an extensive list of other reading materials and websites that will allow you to begin expanding your knowledge of the challenges we'll face and how they may play out. You'll have to make your decision on your living arrangements based on your exploration of many differing views. From my biography earlier in the book you know that I've made the choice to live in the country, ten miles from town and three miles from my nearest neighbor. This may seem a bit extreme to many, and it is partially because this was the off-grid house we could afford. I believe that the nearby village will be an excellent place to live in the future. It is the central meeting place for the many farmers who surround the town as well as for an eclectic variety of urban refugees who bring a varied and creative spirit to the town. Yes, we'll all be driving less, but there is land for the production of biofuels like biodiesel, and once in a while some locals ride their horses into town, not by necessity but by choice. So it won't be a stretch for many to return to some of the older ways.

Do your research, make your Plan, and start scanning those real estate listings for your dream home, wherever it may be.

10   Where to Work

"Meet the new boss. Same as the old boss."
Pete Townshend, "Won't Get Fooled Again"

One thing that should be very clear from our examination of how radically different the future will be is that your choice of job and career is going to be very important in helping you achieve your goal of independence. Having a job at all will be important. Having a job in an industry with some stability or growth potential will be really important. So you have to start analyzing where you work and how likely it is that the company will be around in 5 or 10 or 15 years. The last few months have seen massive upheavals in some of the very industries that have been the backbone of our economy for decades. The five banks that made up the "Wall Street Investment Banks" are now all gone, either bankrupt or absorbed by other institutions. They have shed literally tens of thousands of jobs. The North American car makers are in turmoil and struggling to survive. General Motors, at one time the largest corporation on the planet is bankrupt. It used to be said that "What's good for GM is good for America," so I guess we shouldn't be surprised at the state of the economy in general.

With so much data pointing to the world having hit peak oil and indicating that as we begin to slide down the back side of the curve oil is going to become increasingly and prohibitively expensive for many people, it would appear that the auto sector is a good one to avoid. Most people will continue to drive for some time. Some people will be able to drive regardless of the price of gas because they are wealthy. Some people will switch to electric and plug-in hybrid vehicles, but one thing is clear: as the price of gas rises and we begin to experience shortages we will become a much less mobile society. This means less driving and fewer cars. In the spring of 2008 when oil hit $147/barrel, you could not turn on a TV newscast without a "Pain at the Pump" report. People were really concerned. They were angry and disoriented. As you would expect, the oil companies were called to testify before Congress, but it was a huge waste of time. The price of oil is a world price. The oil companies do not set it. People buying and selling oil and oil contracts throughout the world set the price and they decided oil was very valuable.

The deleveraging of the economic crisis caused the price of oil to crash, but this is a short-term phenomenon. The price will continue an endless march towards hundreds of dollars a barrel—$200, $400, $600. It is such a valuable and essential commodity that it will become very expensive as supply starts to run out. As Matt Simmons likes to point out, $147/barrel oil sounds expensive but some Europeans have been paying prices close to $500 and $600/barrel at the pump. Their governments heavily tax gasoline and diesel to discourage its use and Europeans drive smaller, more fuel-efficient cars and take trains and public transit more than North Americans do. So if you are going to work in the auto industry you need to work for a company that makes extremely fuel-efficient vehicles. Products like plug-in electric hybrids will have a much brighter future than traditional gas-guzzling vehicles. Working for a company that makes passenger railcars would be a better bet because all countries will be quickly rehabilitating their rail service when they realize that the happy motoring days are over and that trains are an incredibly efficient way to move large numbers of people. Once the rail system is improved you will see a positive feedback loop because more people will start using trains, which will pump more money into the system.

If you are in the airline industry now, enjoy the last few months (years) of flying because it is an industry with no future. It is one of the most destructive industries for our atmosphere, and it is premised completely on cheap and abundant oil. Getting a 200-ton 747 jet in the air is a miracle made possible by the incredible energy density of oil-derived fuels like kerosene or jet fuel. As the price of oil inched up towards $147 in the spring of 2008 the airline industry was falling apart. Many companies went bankrupt. Others cut back on flights and everyone began adding extra charges to make up for the rise in the price of fuel. Airlines were getting serious about not sending jets that were half full. Fuel which had been 15 or 20% of an airline's operating costs as late as 2000 was getting closer to 40% during the price spike. Airlines could only take so much water out of the toilet holding tanks to save weight before something was going to give. If the price of oil had remained that high or continued to increase, you would have seen a huge reorganization of the industry, with far fewer carriers and much higher prices. The days of cut-rate air travel are drawing to a close. There will always be some people with enough money to fly, but the industry will be much smaller and employ far fewer people. If you are in that industry your days are numbered.

This holds true for any industry heavily reliant on oil, such as trucking. Right now we move food and products incredible distances, many of them by truck. The economic downturn has seen a decrease in the amount of freight on the roads. As the effect of rising oil prices starts to take hold you'll see a return to much more local economies. Yes, you will have some people able to purchase goods that have traveled a distance, but the model of business we currently have, with massive warehouse stores full of products made 16,000 miles away in China, will face huge challenges as we start running out of oil. We're all going to have far less choice in what we buy and there will be far fewer long distance truckers on the roads.

Think of jobs related to roads and try to stay away from them. This includes warehousing, gas stations, the taxi industry, the RV industry, driving a snowplow, working for a road-building company, and driving a forklift that loads the trucks. Our economy is very transportation focused.

At the same time we're seeing a scaling back of consumer credit and consumer spending in general. People are starting to realize that the credit bender we've been on for a decade was not a good thing. They have less discretionary income to spend so they are using it much more carefully. They are buying fewer frills. Stay away from retail jobs that involve, or industries that make, products that aren't essential. The whole retail industry is going to have to scale back dramatically. People will be making fewer impulse purchases, so all those retail outlets that sell bobbles and bangles that don't really help someone get by are going to be much less in demand. You can extend this to the entertainment industry as well in terms of "products" like live music and even professional sports. People will not have the money to pay exorbitant prices for tickets and drive half way across the city to get to a stadium. The upside of this is that players will finally start making realistic salaries versus the insanity that is professional sport contracts today. Even the NBA needed to borrow $175 million as bailout money for some of its teams to deal with struggling attendance and revenues.

Remember, I'm talking about taking a long-term view on the economy here. For instance, right now with the size of government stimulus packages it might be easier to find a job building roads or fixing bridges than in a green industry. That's all right. Take what you can. Just remember that in the long term we are about to go through some fundamental shifts in where jobs are, so the long-term potential of building highways is not good. If you like construction work, the long-term potential will probably be higher in building and maintaining rail lines, because these are what many of us are going to be moving on.

Government jobs may be one of the most secure long-term places to be. While many of us are loath to consider a government job, if job security is a priority it's one of the best options. If you are going to go after a government job go as high up the food chain as you can. Cities and municipalities are dependent on property taxes for much of their income and obviously with the challenges in the housing market their ability to raise money through taxes will be restricted. State and provincial governments have more flexibility in terms of revenue and more stability. Ultimately the best place will be the federal government, one of the most stable institutions in a democratic society. As the financial meltdown has picked up momentum one of the most popular places to park money is in government bonds and treasury bills because of their security. In fact at the height of some of the panic government bonds actually had a negative interest rate, but investors were still happy to leave their money there because at least they weren't going to lose it.

So what follows is that as long as there is faith in the federal government and people have enough confidence to lend the government money by buying their debt, this is going to be one of the safest places to work. There are those who would suggest that governments could go bankrupt. In other places governments can be overthrown. This has not been our experience in North America and it looks fairly unlikely that it will happen, but nothing is guaranteed in this world. Let's just hope we are able to maintain a strong federal government to keep our society orderly and on track.

The other strong possibility is that as the economic downturn drags on, sooner or later governments are going to have to come to grips with the deficits they are running now. That's how Keynesian economics works. The government should spend and run deficits during economic downturns and then should reduce spending and pay off that debt and run a surplus during the good times. With this year's U.S. federal budget deficit estimated at $3 trillion a well as the $11 or $12 trillion already on the books (not counting unfunded liabilities like Social Security and Medicare), sooner or later the Federal government is going to have to cut back on its spending. Part of that could be the size of the civil service, so there may not be a lot of new jobs available. Many of the existing jobs could be eliminated through attrition, which would be less noticeable and therefore more politically popular.

One of the few bright spots for government jobs will be the military. For many years, especially after the invasion of Iraq, the military had trouble finding recruits. With the economic downturn recruiting is back to where the military wants it because so many people have no other option. Prior to Iraq the military was a pretty safe occupation, but there is obviously a sense of desperation creeping in for some people to consider this option today.

There will be good potential for employment in the security business. As the world gets more dangerous people will be willing to pay more for security. Jobs in the security business will include everything from security guard to alarm system installer to bodyguard. So it may not be a bad idea to stay in good shape and be comfortable with firearms. There are some security jobs, for example, monitoring cameras for alarm companies, where it's unlikely that you'll be in high demand to chase bad guys. But if you do think this is an industry you're interested in you should start taking martial arts and programs that give you more confidence in conflict resolution situations. Lots of former military people pursue these careers because they've received this training in their previous jobs and were paid to do it. There is that stint, though, where they risked their lives for their country, so think about that if you're considering the military as your entry into the security trade.

If you can't work or don't want to work for the government, where else should you be looking? People are going to be getting back to basics, so you need to be in an industry that caters to basic human needs: food, shelter, clothing, warmth, a few luxuries. Food is going to be key and it is one you may want to zero in on. I'm not talking about developing a line of exclusive balsamic vinegar and Gorgonzola marinades. I believe the days are numbered for these types of products. People simply aren't going to be able to afford them and are going to be returning to a much more basic diet. The days of fancy-shmancy are drawing to a close. The days of just being able to afford a decent meal will soon be upon us, so you can set your sights much lower in this regard.

This is good because it's not going to require such advanced skills. What you're going to need to be able to do is to grow or sell food for people. If you're going to grow enough food to support more than yourself, you're probably going to need more room than your suburban lot. So you'll need to find some land to purchase or rent. This also may become part of your long-term plan, to move to a rural location with enough room to grow food for market. This is a good strategy, but don't wait until you get to the country to start honing your gardening skills. As you'll read later, you should start now with your city lot and get growing!

Selling food someone else has grown may be a good idea as well. This is what grocery stores have done so well over the last decades. The difference is that their model is going to be harder to maintain. They are too big and their inventory comes from too far away. This model is going to stop working as energy prices skyrocket. So as we power down and downscale, smaller stores providing food will become the norm. And it may be a more specialized store that might make sense as people localize. You might decide to specialize in producing eggs to supply local stores. Or you might want to open a bakery. Yes, we've all become used to industrial food, baked in ovens hours away, but many of us will begin returning to sourcing our food locally, and a bakery is a great place to start. The same advice applies here; don't suddenly decide you're going to open a bakery. Be a good baker first. I believe I can cook fairly well, but I am bread and dough challenged. I can never get the yeast the right temperature and my baked goods are legendary for being hockey puck substitutes. This is a skill I hope to work on in the future but I would not make it operating a bakery.

The more logical cog in the food chain will be growing it yourself and selling it yourself. Later I'll discuss some of the techniques like Community Supported Agriculture that may help you ensure a market for what you produce. In the future, if you have land and you can grow food you will be financially secure. As the high price of oil results in food being moved over shorter distances, locally grown food will be increasingly in demand. As we all spend a higher percentage of our incomes on food, the people that grow it will be better able to earn a living. Another upside of the downward trend in world energy reserves is that large-scale farming will become less viable. And the cheap nitrogen-based fertilizers made from natural gas will disappear, along with the many other petroleum inputs into the current agribusiness that large farms require. At the same time, there will be a financial crisis and credit will no longer be readily available. The scale of financing which some of these large farms require is staggering, so as capital becomes more scarce farms will find it increasingly difficult to operate on the same scale.

I think a lot of farmers will be much happier working at a more sustainable level. There will be less stress for them because the scale of the operation will be more manageable. I'll bet if you asked many farmers with huge operations whether they would scale back if they could earn the same income farming half or a quarter as much acreage, they would say yes. Farmers are not greedy business people. They have a passion for producing food and for being stewards of their land, and the economics of the modern agribusiness has forced them to go big or go home. I think older farmers especially would welcome a scaling back.

When you look at the upheaval in Cuba when Russia collapsed and Cuba could no longer source fossil fuel inputs for agriculture, the country turned to its seasoned and experienced farmers and said, "How did you use to do it?" They needed to relearn some of those old skills that had been lost. So farming will return to a smaller more humane scale here as well, and it will present many opportunities for someone wanting to participate. Being able to grow food will be invaluable not only for your food independence but for your financial independence as well.

While "green jobs" have become a real catchphrase of the time I believe that they do offer you real potential in the future. If you are currently involved in the manufacturing or industrial sector, even if you do have a job it is a good idea to start looking at moving to a company involved in producing renewable-energy products. The demand for large-scale wind turbines and solar panels is increasing dramatically, and I believe that as we run out of cheap energy to ship products from countries with lower production costs, there will be great potential for North American manufacturers to supply our own market. Currently many large wind turbines come from Europe and solar panels from Japan, Germany, and other countries. If our leaders are talking about green jobs and putting taxpayer money into creating these opportunities I believe there is a subtle expectation that local jobs will be created in the production of those products.

Manufacturing the products will be one area to explore, and selling and installing them will be another. There has been explosive growth in the renewable energy industry as more and more homeowners and businesses want this equipment on their roofs. *Home Power* magazine (www.

homepower.com), which was the original home-based renewable energy magazine, started out as a very thin, rough, black and white newsprint product. Passionate people in a non-existent industry wrote it. Now it's a large, glossy, beautifully produced showpiece filled with ads from endless companies in the industry. I like to just read it and marvel at the variety of products and suppliers available today. *Home Power* has now spun off a new magazine called *SolarPro*. It ratchets up the industry to the next level and is targeted towards people who actually sell and install solar equipment. How great is that! The business is now so big it has its own magazine. And there are many other new renewable energy magazines all the time.

A variety of educational institutions are now offering renewable energy training and sustainability courses at a diploma and degree level. There are also many private institutions like Solar Energy International (www. solarenergy.org), which has been providing training at different locations for a decade and a half. SEI programs allow you to obtain certification in many areas of the installation industry.

In the ten years I have lived off the grid I have seen a remarkable evolution in the industry. The people we originally purchased our equipment from were almost reluctant to talk to us at first because so many people came in and wasted their time. Everyone wants to save the world but no one wants to pay for it, so they had to support themselves by selling wood stoves. They now have a profitable renewable energy business and continue to hire new staff all the time. Another dealer who started out doing energy audits is growing by leaps and bounds. Our nearest city of 100,000 people has at least five renewable energy dealers now and they all seem to have enough work to stay in business. I think this industry has a very bright future, literally! Many people will be installing these systems as backup against grid failures, and many others will do it to save the planet. The end result is the same: jobs.

I also think that if you have an aptitude for fixing things you will do well in the future. People will be hanging onto "things" for much longer than they do now. Whether it is cars or washing machines, the dizzy credit frenzy is over and most people (especially readers of this book) are going to start living within their means. If they don't have the money to replace that broken washing machine, they are going to need to fix it. The same will hold true for cars. As vehicles increased in cost over the years many people switched to leasing them to have a fairly new car that didn't break down that often. Now not only will many of us not have the income to lease a car, which can add even more expense to car ownership, but many of the leasing avenues will dry up as credit evaporates and car companies close their leasing divisions.

Fixing cars will be one area where there will be work available for some time. I obviously believe that peak oil will limit the prospects of this industry, but as long as we can get gas I believe some North Americans will be willing to devote a much larger portion of their income and assets to staying mobile. Some people will have the income required to keep a car running. And as the economic crisis hits every part of the world, there is the possibility that there will be enough demand destruction for oil that the remaining reserves will last longer as we make our way down the back side of the peak oil curve. It wouldn't hurt to be a diesel mechanic as well. Diesel engines operate differently from gas engines and require some unique skills. They can run on a fuel made from waste vegetable oil called biodiesel. Diesel engines can also be retrofitted to run on pure vegetable oil. And I believe governments will make diesel fuel a priority for some truck transport (and military vehicles) and farm equipment. So being able to work on a diesel engine will make a lot of sense.

I think the hardest concept for most of us will be the concept of moving away from this paperless, computer-controlled work environment where we spend the day moving pixels of data around on our computer and across the country from our desks. Yes, that will continue, but it may represent a much smaller percentage of the economy. There will be banks and insurance companies and governments that require people to do this, but there will be far fewer people employed in this way. I think the economy will turn increasingly to jobs where people make "things" or provide "services" on a more personal level, where you know the person you're trading with. The transaction may involve cash or it may involve barter, but it will be more local and more hands on. If you've always liked crafts and working with your hands maybe it's time to get some property and learn to keep bees. Natural sweeteners like honey will be very popular when sugar from tropical countries gets expensive. With beeswax you can make candles and a whole variety of natural skin care products. I emphasize "natural" because right now to read and understand the label of a bottle of hand cream you need to be a cross between a petroleum engineer and a biochemist. When oil gets scarce and expensive, having a local supplier of natural products that often do a better job will be a luxury many people will be happy to have.

With the housing boom of the last decade a huge number of workers were attracted to that trade and now find themselves without work. The economy is about to become more focused on improving existing assets rather than building new ones. So you should be able to take the skills you learned in a trade and move them more towards renovations. As economies become more local people will want to be closer together, and you will find areas of town where buildings that have been vacant or left to deteriorate are now coming back in fashion because they are within walking distance of amenities and need to be renovated. If you're an electrician, why not expand your skill set from traditional grid wiring to include installing renewable energy systems? This is a growth area. If you are a plumber, why not take a course on solar thermal systems and begin installing solar domestic hot water systems? Since they tie into existing hot water plumbing systems you are the natural person to be installing them. If you are concerned about having the capital to go out on your own, find a solar company that is marketing these systems and offer to contract the installations. They may be just as happy to not have to put someone on their payroll to do this. Or why not approach the local utility and suggest that it offer to lease the systems to their customers and have you do the installs? There will be huge potential in providing equipment that allows homeowners to become more independent and less at the mercy of rising energy costs.

Some cities are relaxing their restrictions on keeping urban animals and are allowing city dwellers to keep animals like chickens. If you love animals and free eggs, maybe you should invite several chickens to take up residence in the coop you build in the backyard and see how you get along. If you enjoy their company, perhaps acquiring a small acreage to raise chickens on a larger scale could be part of your five-year plan. Chickens today are raised on an unsustainable megascale and the energy inputs are enormous. At a recent local sustainability symposium where I gave the keynote address I spoke to a farmer who was trying to figure out how to stay in business. He had installed a heat pump to keep his home warm but the propane he required to keep his massive chicken operation heated was astronomical. I reminded him that propane is the most expensive energy source, coming from the top of a barrel of crude that is "cracked" at a refinery into various components like gas and diesel. There is no long-term viability of this model of farming. A smaller-scale version where you heat a smaller barn with wood cut sustainably from your property does work. It's better for the planet, for the welfare of the animals, and for supporting your family sustainably. As we all start burning more calories to replace the ones we won't be displacing with fossil fuels, there will be days when we won't mind a big breakfast with scrambled eggs to get us going. If today's the day we're riding two hours to the nearest city to buy a slip ring for the wind turbine or finally clearing that patch of brush to make the vegetable garden a little bigger, we'll be happy to barter for some local eggs.

I see nothing but good coming from this. I see too many people sitting at desks doing jobs they hate that provide them with no joy. These jobs suck your soul away and leave you drained and without energy by the end of the day. Fixing a tractor someone desperately needs to get hay in, providing someone with honey for the morning tea and toast they love, providing the local diner with eggs that your neighbors eat to give them energy to get through the day, these are wonderful things. These will fill your soul with joy and bring real meaning to your life. That wagon full of corn that you've grown and that you know will provide your neighbors with eating pleasure as well as caloric energy will give you a far greater sense of accomplishment than any paycheck you might have received for importing plastic rocks from afar.

This economic downturn presents incredible opportunity for many of us to create a much more meaningful life for ourselves, but finding opportunity in chaos can be hard. I have no crystal ball to point you in a guaranteed direction for earning an income in the future. Getting involved with your local community is probably one of the best places to start to look. Perhaps your part of town is about to lose a large grocery store, which presents an opportunity for someone to open a smaller outlet with lower overhead. People will patronize it because it's part of the community and close by and so reduces expensive driving. Perhaps you're in a town where the power grid is becoming less reliable and you sense at meetings a growing frustration with the large electrical utility. Time to start installing some solar panels for your neighbors. Perhaps we've had another run-up in gas prices and you realize that your neighborhood doesn't have a bike shop and yet lots of people are now riding to work and cycling more. Maybe that retail store four streets over that has been vacant for a year or two has a landlord willing to negotiate a reasonable rent until you can get your bike shop up and running. Selling and fixing bikes will probably become a growth industry.

Or maybe you've been baking bread and desserts for your family and friends and they've been raving about them for years. Maybe now is the time to start selling your wares and once you have established a regular clientele find a retail space central to people in your neighborhood who love freshly baked quality bread. Perhaps by linking up with a local farmer who grows grain and mills his own flour you will can market yourself as "The 100-Mile Bakery" and develop a clientele who want to eat sustainably and enjoy the healthiest food they'll ever find.

Our economy is going to return to being more local. It's going to be based on rehabilitating existing infrastructure rather than building new buildings. It's going to mean that you will be more likely to know the person you are exchanging money or labor with and if you do a good job they will recommend you to others. You will not get job "performance reviews" by some manager who spends next to no time with you. You will get recommendations based on the quality of your work. The better you do, the more people will recommend you and the more money you'll make. This is the finest attribute of a capitalist system. It rewards quality and innovation. Right now with your current industry in decline you may be forced to evaluate how you earn your income, but necessity is the mother of invention. The new job or business that you create will be far more rewarding and will provide much greater satisfaction. Can you imagine lying in bed every morning feeling the excitement of contemplating the opportunities that today's work presents? Can you imagine not dreading that walk into the office? Can you imagine not being obsessed with your next holidays and spending days off working on things that improve your skills and your business?

This is the great opportunity of this economic dislocation that a democratic and capitalist system provides. There will always be opportunity. You must now take a long-term look and determine where you think these will be. Now is also the time to do a personal self-examination and decide what you do well and what your likes and dislikes are. Yes, if you've lost your job and have a mortgage take any work you can. But when you have a spare moment start analyzing what you'd like to be doing in the future and begin building the skills and strengths you'll need for it. You might as well pick something you're going to enjoy. View this economic upheaval as one great big chance for you to get your life and career on a track that will ultimately bring you satisfaction and happiness. The glass is half full!

11   Heating

When I come home cold and tired
It's good to warm my bones beside the fire
Pink Floyd, "Time," Dark Side of the Moon

Staying warm is a good thing.
It's a very basic thing. It's very low on Maslow's hierarchy of needs, but it's a big priority for many North Americans. It is going to become a big challenge for many of us as the sources of heat that we've relied on become less dependable.

Natural Gas
While it's likely that the world has hit peak oil, it's even more likely that North America has hit peak natural gas. In his book *High Noon for Natural Gas* Julian Darley provides a thorough investigation of where the continent is with natural gas, and it doesn't look good. Natural gas is a continental fuel. It's a gas, so when you drill a well you have to get it into a pipeline. Once it's there you can move it around, but there are no gas pipelines coming to North America from across bodies of water. What happens in North America stays in North America when it comes to natural gas.

The United States appears to have peaked in its natural gas production. Canada, which supplies 15% of U.S. natural gas, has peaked as well. Canadians drill more and more natural gas wells and discover less and less natural gas. And while 15% may not sound like a lot of natural gas to Americans, it's 55% of Canada's natural gas. When the two governments signed the North American Free Trade Agreement, Canadians consented to keep sending that percentage of natural gas even if they are running out and need it themselves.

The other challenge with Canadian natural gas is the tar sands. Extracting the tar-like bitumen, which can be processed into low-grade crude oil, requires massive amounts of energy in the form of heat to separate the bitumen from the sand that it's combined with. Most of that heat comes from natural gas. Right now Alberta's tar sands mega projects use more natural gas than would heat four million North American homes. This is a very high-quality fuel which is best used for things like home heating, and instead it's being used to produce a low-quality transportation fuel that must be mixed with high quality crude oil before it can power any vehicle.

So where are we going to get our natural gas as supplies dwindle? LNG or "liquid natural gas" is what many are hoping will alleviate the challenge of natural gas shortages. So what is LNG? We get LNG from countries that have extra natural gas that they don't need. Countries like Iran don't have the same home-heating requirements as many North American homes, so we get LNG from them. Many of the countries we get LNG from are not necessarily considered friendly to this part of the world. Russia and Nigeria also supply LNG to much of the world.

To produce LNG the processor turns the natural gas into a liquid. You do this by chilling it to 163°C below zero. This process uses a massive amount of energy. Once you get it that cold, you put it into LNG tankers to ship to countries that want it. LNG is very explosive, so you need to put it in special tanks where it can't escape. In fact in a recent ad in *The Economist*, one of the companies that makes the holding tanks for LNG was advertising for employees. The ad states: "LNG is volatile. If not stored at 163°C below zero in the most secure, leak-proof tanks, it explodes violently." This is the explanation this company uses to attract employees. If this is the upside of working there, I'd hate to see the downside.

A huge explosion at a natural gas plant in Algeria in 2004 killed 26 workers, injured hundreds of others, and caused widespread destruction. Not surprisingly, very few people want terminals built near their homes to unload the LNG after it has been shipped across oceans to North America. The LNG is piped to holding tanks, and large amounts of energy are used to turn it from a liquid back to a gas to put into pipelines to be shipped across the continent.

A wayward spark of static electricity, a human error made during unloading, or a well-aimed rocket-propelled grenade could lead to a Hiroshima-sized explosion. One can't really blame anyone for not wanting one of these too close to their home.

California, which uses a large amount of natural gas for power generation, discovered this when it realized it had to start importing large quantities of LNG. Its solution was to build the plants in the Baja peninsula in Mexico and ship the gas up from there, because there was less opposition to the unloading terminals in Mexico.

So we have a continent that appears to have hit or nearly hit its maximum amount of natural gas and also has a physical limitation on how much liquid natural gas can be imported to try and make up the difference. The result? Higher natural gas prices—potentially outrageously high natural gas prices. And shortages. There is a possibility that if the supply of natural gas deteriorates there will be some areas of the continent that will experience shortages of natural gas. Anyone who has experienced frigid winter weather knows what a bone-chilling prospect that is.

So if you heat with natural gas right now, you need to start thinking about a backup system or switch to another heating source altogether.

Home Heating Oil

How about switching to home heating oil? Lots of people are talking about it.

The reality is that home heating oil provides the same dilemma that natural gas does. Heating oil is a carbon-based, non-renewable fossil fuel. As discussed in Part I, it appears that the world has hit peak oil and we will have to deal with ever-decreasing amounts of oil. Home heating oil is basically diesel fuel, which is produced when a barrel of crude is processed. Some of the barrel ends up as gasoline, some as other petroleum-based products such as jet fuel and asphalt, and some as diesel fuel or home heating oil.

So if you heat with oil, you'll be competing with trucks that use diesel to haul goods around the continent, municipalities that use trucks to plow snow and repair electricity networks, trains that burn diesel to move people and goods, and a whole host of other competitors that use oil, mostly in the transportation sector. Some would argue that the economic downturn will greatly reduce the volume of goods being shipped. But the reality is that we have developed North America to be very transportation-dependent, from the movement of goods across the country to the fact that we live in car-dependent suburbs that require driving to jobs and shopping.

So home heating oil does not offer any long-term security of supply.

And price will be very volatile. With the plunging of commodity prices it has fallen to more manageable levels. But oil is an incredibly valuable resource and we were given a one-time endowment of it. For the last hundred years we have been pumping and burning the remains of plants that existed on the planet millions of years ago and through a miraculous process of heat and pressure gave us this incredibly powerful gift. But the endowment is starting to run out and there is nothing readily available to replace it. So there is no long-term future in heating with oil.

Now if you owned a restaurant or a french fry truck and had access to large quantities of vegetable oil from your deep fryer, then you would have an option, and that's biodiesel. That would require you to use "methanol", a natural-gas-based chemical which precipitates the glycerin out from the oil, leaving "fatty acid methyl esters" or biodiesel, which you could burn in your oil furnace. You would have to keep the oil inside so it didn't freeze in cold weather, but if you have access to large quantities of vegetable oil this is an option. You can learn more about biodiesel in William Kemp's *Biodiesel Basics and Beyond*.

Electricity

Some people heat their homes with electricity. This can take a variety of forms such as an electric furnace, baseboard heaters, or on-demand hot water heaters or boilers that heat water that runs through radiators or in-floor radiant heating pipes.

Depending on where you live, this may be a cost-effective way to heat your home. At least it may be less expensive than fossil fuels like gas and oil. But electricity can be expensive to make, so its price, while reasonable today, is by no means guaranteed.

Electrical utilities have not been investing in new generation or maintaining their existing generating facilities. So most North American electrical utilities have to make massive investments in the future to keep the lights on. They have also not been investing in the infrastructure of transmission lines that bring the electricity to your house.

The North American Electric Reliability Corporation, an organization funded by the power industry has been trying to avoid another large-scale blackout.

The US power grid - three interconnected grids made up of 3,500 utilities serving 283 million people - still hangs together by a thread, and its dilapidated state is perhaps one of the greatest threats to homeland security, according to Bruce deGrazia, the president of Global Homeland Security Advisors and a former assistant deputy undersecretary for the Department of Defense, who spoke at an electricity industry conference in Shepherdstown, Virginia.[1]

After the blackout on August 14, 2003, which left 50 million North Americans without electricity, President Bush called for spending of up to $100 billion to fix the system. But the money that would have funded a reliable power grid was spent on the wars in Iraq and Afghanistan.

So the grid is going to become strained because demand for electricity grows and the infrastructure to deliver it has not been upgraded. Plants that supply the bulk of our electricity are getting older and will become less reliable and more costly to maintain. And someone who cares about their footprint on the planet should not be heating with electricity when so much of it is produced by coal.

In the United States, 50% of electricity is generated with coal. This is a very dirty way to generate electricity, and even though you don't produce any CO_2 at your house when you turn on your electric range to cook dinner, in some states that are 100% coal powered you are having an extremely negative effect on the health of the planet through your use of electricity.

Those in districts that have nuclear power have another conundrum. Yes, nuclear power doesn't generate any significant greenhouse gases; it does create a legacy of nuclear waste. No government in the world has been able to devise a plan to safely store spent nuclear fuel during the thousands of years it will remain lethal. When your electricity comes from nuclear power, you contribute to this nuclear waste, which is also the material used to make nuclear weapons. There are lots of downsides to the nuclear option for power generation.

So the traditional sources—natural gas, home heating oil, and electricity—don't have a good future as potential heat sources for your home. Remember that gas and oil and the coal used for power generation are non-renewable fossil fuels. When you burn them you extract a carbon-based fuel that has been trapped beneath the surface of the planet, not doing any harm to the atmosphere, and you release CO_2. Carbon dioxide is the leading cause of global climate change. Even if you burn natural gas, which is considered "cleaner" than many other fuels, the reality is you are still contributing to climate change in a fairly big way.

Wouldn't be nice if you could find a way to heat that offered more security and didn't do as much damage to the planet? There are two readily available forms of heating that use renewable energy: geothermal and wood.

Geothermal

When the sun hits the earth, the ground absorbs some of its heat energy. If you were to dig down five or six feet, you'd find the earth stays a fairly consistent 40 F. (5 C.) to 50 F. (10 C.) year round, regardless of the conditions six feet above. If you circulate liquid through a series of pipes in the ground, you can use that heat for your home and, best of all, run the system in reverse and use it to cool your house in the summer. This is an unlimited energy source; you merely have to use some energy to bring that stored solar power into your home.

With a geothermal system, or "heat pump" as some people call them, you pay roughly 25% of your home's heating and cooling costs while the rest comes free from your backyard. Here's an example: for every 1W of electricity you purchase from the electricity grid (to power the pump that moves the liquid) you get 3W of heat energy from the ground, giving you a total of 4W of heating or cooling for your home. Yes, you do still require some electricity, but this is much more efficient than using only electricity in resistive heating like baseboard heaters.

Geothermal systems require a loop to run fluid through the ground to extract heat. These can be horizontal loops, vertical loops, or pond or lake loops. In a horizontal loop, trenches are dug six feet deep and pipe is laid in loop circuits. In a vertical loop a series of holes are drilled in the ground and pipe is placed vertically in the boreholes. In a pond or lake loop, the pipe is placed in the lake or pond and anchored offshore where it can't freeze.

While it may seem counterintuitive that the cold water of a lake in the winter could heat your house, the trick with a geothermal system is that it uses a compressor to operate a refrigeration cycle that extracts heat energy from the ground or lake and upgrades its temperature to a level high enough for space heating. In the summer, the process is reversed to provide air conditioning. The unit may also provide a portion of your domestic hot water.

The cost of a geothermal system may be more than that of a traditional fossil-fuel-based heating system like a natural-gas forced-air furnace, but this is not how to evaluate the purchase. That furnace will require you to continue purchasing a nonrenewable fossil fuel that most indicators suggest will be increasing in cost dramatically in the future. You should not use the purchase price but the lifecycle cost of the heating system, what it's going to cost to operate over its lifetime. Also remember that with a geothermal system you are purchasing both heating and cooling. If you compare the cost with that of a new high-efficiency furnace and a central air conditioning system, the geothermal system will be very competitive without the long-term commitment to the continued purchase of fossil fuel.

Wood

There are lots of misconceptions about wood heat. Some people think it is very polluting. Others think you can't heat a house with wood because they've experienced a decorative fireplace in the living room that actually sucks heat up the chimney rather than delivering it to the house. But a modern EPA-certified wood stove will burn very cleanly and, if situated correctly in your house, do an excellent job of heating it.

In the 1970s, during the first energy crisis, there was a renewed interest in wood burning, and the market was flooded with a variety of wood stoves. While many looked great and did a reasonable job of heating, the technology has continued to improve and evolve to the point where wood stoves are an excellent alternative to other heating systems.

Wood stoves are one of the most environmentally sustainable ways to heat because they use a renewable fuel—wood. When a tree is growing it uses photosynthesis to store carbon, extracting carbon dioxide from the air. When the tree dies it releases the carbon it stored while growing, making it "carbon neutral." If a tree dies in the forest and falls to the forest floor, it will release the same amount of carbon that it absorbed during it's life, and it will release the same amount of heat energy as it would if you burned it in a in proper EPA-certified heating appliance. You're just speeding up the release of that heat when you burn it..

The two main types of wood stoves are catalytic and noncatalytic. A catalytic wood stove uses a catalytic combustor similar to the catalytic converter which makes your car pollute less. In this wood stove the wood is burned in a primary combustion chamber and the smoke is then diverted through the combustor, which is usually a honeycomb made from refractory material that gets very hot. As the smoke passes through the honeycomb, any unburned gases or particulate gets burned off, making the remaining exhaust very clean.

In a noncatalytic wood stove you have just one burn chamber, but you bring oxygen into it so that after the primary combustion there is a secondary combustion to clean up the exhaust before it goes up the chimney. I have had both wood stoves and much prefer the non-catalytic. Our catalytic wood stove required us to put it into an airtight mode after it got hot enough. This made the stove more labour intensive. It also required us to purchase the catalytic combustor on a regular basis for $200. I was told they should last close to five years but I rarely got more than two years out of them. The other downside to our catalytic stove was how many places air could get in. There were numerous gaskets that had to be replaced to prevent outside air from getting into the stove and affecting its performance. If too much air gets into the stove it burns the wood faster and therefore requires loading more often and goes through more wood. This stove was time consuming and expensive to maintain and never worked as advertised.

Our new Pacific Energy wood stove is exceptional. It has only one gasket to be maintained, which is on the front and only door. It burns exceptionally cleanly and when loaded with hardwood burns for 10 to 12 hours. We could rarely get more than 6 hours from our catalytic woodstove. Now we can load it up at night, have a sound sleep, and get up to a bed of glowing coals. We just toss in some more wood and have it roaring again in minutes. It's quite brilliant.

People probably assume that heating with wood is an option only if you live in the country. There are companies that will deliver firewood to cities, so it may be worth tracking some down and getting a handle on costs. There is also a huge untapped source of firewood in urban areas that comes from people trimming and cutting down trees on their property. This is often left on the lawn for someone to grab. A simple knock on the door when you see the tree being cut or notice a pile of unclaimed firewood often gets a response of "That would be great; I'd be so glad to see someone haul it away."

There are lots of trailers you can buy to haul this wood home. And cities have made sourcing this wood even easier by having pickups of "organic" materials periodically. This is often in the spring when people are trimming things back. I'm amazed at the size of some of the wood I see people bundling up to truck away. It's yours for the taking and it offers a huge untapped heating source for your home.

You may not have a huge choice about the quality of the wood, but if it's free, who cares. Ideally you want hardwoods like maple and oak, which will burn for longer and have more heat potential. Softwoods like poplar or birch won't produce as long a burn, so you'll have to burn more of it. If you have both in your woodshed, you'll like the softwoods for the fall and spring when it's not brutally cold and hardwoods for the coldest months of winter. If all you have access to is lower-grade softwoods you'll just have to budget more space for storage since you'll need more. Many areas that don't have as much hardwood have more temperate climates, which helps.

If you are harvesting your own firewood, the size of woodlot you need depends on a number of things: how big your house is, how well insulated it is, whether you like to walk around in a t-shirt in January or a nice thick sweater and wool socks. It also depends on what type of wood you have on your property and how well managed it is.

A fairly typical North American house needs around 20 acres to produce a sustainable woodlot. This assumes a good mix of trees and selective harvesting, meaning that you cut larger more mature trees and allow the light to get to the smaller and lower trees so that they grow more vigorously to fill in the gaps. To properly thin the woodlot you should be looking for poorly formed, crooked, insect-damaged, dead, and undesirable trees. While it doesn't have to look like a suburban lawn, removing trees that don't have the long-term potential to be large and vigorous trees makes for a healthier woodlot that can provide an endless supply of heat for your home.

Sound woodlot management also assumes you are not harvesting with a huge machine which kills all that lower growth. You can hitch a horse up to lengths of trees and haul them out of the woods and up to the house.

I use a large plastic sled which was designed to be towed behind a snowmobile but is very light and rugged. Once the tree is down I cut it into longer lengths comprising two or three fire-sized logs. I pull this out to the house or to where I can get to it with the truck in the spring. I'm careful not to disturb too many of the smaller trees and seedlings on the ground. Once I get these longer lengths back to the house I cut them with either the gas-powered chain saw or an electric chain saw. There is nothing more inspiring than cutting a carbon-neutral heating source like firewood with a solar powered chain saw.

I have yet to figure out how to get the electric saw out into the woods with me, so I still use the gas-powered chain saw. The unit I have has a catalytic converter to help reduce emissions. If gasoline becomes prohibitively expensive, I'll cut the tree by hand, drag it out with a horse, and cut it into smaller lengths with the electric chain saw.

Pellet Stoves

Some people have discovered heating with wood pellets in a pellet stove and it's an excellent option as well. Pellets are made by compressing wood waste from sawmills into smaller pellets that look like rabbit food. They are fed into the burn chamber of the pellet stove by an auger. These stoves require electricity to power the auger that delivers pellets to the firebox and the fan that blows air out of the stove, so make sure you have a backup source of electricity to keep your home warm during a power disruption.

Pellet stoves and pellet sales seem to be very irrational. Every time there is a spike in natural gas prices it is hard to buy a pellet stove because they are snapped up. Then the next year if natural gas prices settle down pellet-stove sellers will have a warehouse full of unsold stoves. The pellets themselves are also prone to the vagaries of the market. When they are abundant, the price is very affordable. When they are in short supply the price rises. Much of the wood waste that they are made from comes from the mills that make lumber for new homes. With the downturn in the North American housing industry these mills are not creating as much waste, so the supply of pellets has been severely restricted in many parts of North America. People who bought pellet stoves thinking they'd removed themselves from the ups and downs of the natural gas market are coming to realize there really is no shelter from the fossil-fuel-depletion storm.

As the price of all energy rises, consumers will fuel switch and they'll do it often and very creatively. If natural gas gets too expensive, they'll buy pellet stoves. If wood pellets get too expensive, they'll plug in space heaters and use more electricity. The reality is that it will be difficult to make a long-term plan for heating your home unless you have access to your own woodlot or heat with a ground-source heat pump and have your own significant renewable energy system to generate the electricity you'll need to move the liquids through the heat pump to recover the heat. Heating with oil, natural gas, electricity, or pellets or firewood that you purchase from someone else will become a wild ride in the future, and the direction of the prices you'll be riding will be constantly upwards.

If you currently live in an urban area where you heat with natural gas, it's a good idea to have a "Plan B" heating source. Perhaps it's to make sure you have some space heaters to keep rooms you spend most of your time in comfortable. And if you believe as I do that the electricity grid will become less dependable after years of underinvestment in new generation and infrastructure, it wouldn't hurt to have a "Plan C." That might be a wood stove in the basement for those emergencies. Every time you take a trip to the cottage bring half a trunk of firewood home, and when that neighbor down the street finally decides to take down the silver maple that keeps dropping branches on his new car, make sure you're there with your wheelbarrow to haul away the fruits of his labor. It's free heat, and if you burn it in an EPA-certified wood stove, it's carbon neutral and an excellent way to heat.

Heating your home in the future is going to be a challenge. It's going to require more of your time to think about how you're going to do it, and it's going to require more of your time and income to accomplish. Cutting your own firewood will be cost effective, but it will require more of the time you formerly used to earn an income or watch football games on Sunday. If you now cancel your health club membership because you're getting such a great workout cutting, hauling, and splitting firewood, all the better.

What's important is that you realize that we are coming into a time of instability and rapidly increasing prices for home heating. Analyze where you live and what resources are near at hand. Do you live near lots of forests? Are you close to sources of natural gas? Has your local utility been investing in new electricity generation so that a reliance on it as your primary heating source may be realistic? Pick what looks to be most cost effective over the long haul, and then spend more money on installing your "Plan B" backup heating source. And if you're like me, make your family think you're crazy and have that "Plan C" system ready to go. There's nothing worse than being cold, and with the knowledge you now have about how uncertain the future will be, you have no excuse to ever be cold.

12   Fuel/Energy in Your Home

And God said, 'Let there be light' and there was light, but the Electricity Board said He would have to wait until Thursday to be connected.
Spike Milligan, Irish comedian

For some people reading this book, heating will be less of a concern than cooling their homes. According to the US Department of Energy the average North American home uses 52% of its onsite energy in heating, 22% for appliances, 17% for hot water, 4% for refrigeration, and 4% for cooling. If you live in Arizona obviously your cooling requirements will be much higher, but this is an average.

While some of the energy requirements other than heating, like hot water, will be supplied by natural gas, many will be provided by electricity, which, like all large-scale energy sources is going to be challenged in the future. The fuel used to make electricity, like natural gas, coal, and uranium for nuclear power plants, will become more expensive and harder to come by. Utilities across North America have not been installing enough new generation capacity to deal with the increased demand. This is partially the "BANANA" effect. We've gone from NIMBY (Not In My BackYard) to Build Absolutely Nothing Anywhere Near Anything (or Anyone). This has made it more difficult to construct and commission new generation.

Utilities have not been investing properly in their infrastructure. The transmission lines, transformers, and all the equipment needed to get electricity from the power plant to your home is like anything mechanical and it needs maintenance. And as demand has grown the infrastructure required to move higher volumes of electricity hasn't been upgraded fast enough to keep up, so the system is taxed.

In the blackout of August 14, 2003, 50 million people in the northern US and southern Canada discovered just how problematic an interconnected, complex system could really be. The assumption at the time was that it shouldn't happen again. President George Bush is on record as saying that we needed to invest $100 billion in electricity infrastructure. But one thing led to another and the money got sucked into Iraq and the problem wasn't really addressed. The electricity grid needs a massive investment in infrastructure at a time when government budgets are strained, revenue is dropping from reduced economic activity and the credit crunch has restricted some traditional sources of large project financing.

So you should start assuming that electricity is going to be more expensive—a reason to use less—and that it's going to be less reliable. This should be all the incentive you need to start making yourself more independent when it comes to electricity. Lots of people are already inspired to put up solar panels to do the right thing for the planet, but there are important steps to take before you put up solar panels. Using the guide created by William Kemp in his book *$mart Power*, you need to follow the "eco-nomic" approach to adopting renewable energy. And that means starting with "nega-watts" or energy saving first, which you accomplish through energy efficiency.

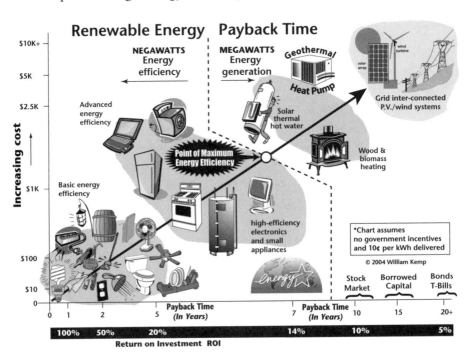

Most people want to start putting up solar panels to generate electricity, but if you follow this chart you'll see that you should really start with simple, inexpensive solutions to reduce your energy consumption first. It's less sexy and not as easy to brag to your neighbors about, but saving energy has a much faster payback than generating energy. That's why one section of the chart is labeled "negawatts," a term coined by Amory Lovins and The Rocky Mountain Institute to represent saved energy.

As you think about those items that give you the fastest payback, a light bulb should appear above your head. It should be a compact fluorescent light bulb (CFL). The traditional incandescent light bulbs that use a hundred-year-old technology are much better producers of heat than light. You can tell this by examining your kid's Easy-Bake Oven. It uses a 100W light bulb to bake cakes! And we've continued to use these bulbs as central air conditioning has become more common, so you have this double whammy for your house. You're using electricity to cool your house, then you're using more electricity, very inefficiently, to light it, and then you have to use more electricity to remove that extra heat you've created with incandescent bulbs. Countries like Australia and Canada have begun a phase out of these older incandescent bulbs with an eventual move to a complete ban, so the writing is on the wall.

Not only are CFLs five times more efficient than incandescent, they last ten times longer. So even though the initial purchase price is higher than the price of incandescents, over the life of the bulb you'll save a lot of money

The per-hour cost of energy for CFLs is cheaper. Let's say you use a 26W CFL to replace a 75W incandescent, resulting in a difference of 49 watts per hour of usage in favor of the CFL. If you use that bulb for four hours a day and your energy cost is ten cents per kilowatt-hour, that saves you $7.15 over 12 months. An average home has 20 bulbs, so you save almost $150 a year. Since the CFL bulbs should last ten years, that's a $1,500 savings over ten years. And many people pay more than ten cents per kilowatt-hour once you factor in the additional charges that now appear on many electricity bills.

So you need to start replacing your light bulbs. You should experiment with different brands to find the ones you like. A good-quality bulb will come with a five- to seven-year warranty, and I know there are a number of CFLs in our home that were there when we bought it more than ten years ago. So if you are one of those incredibly organized people, save those receipts in case anything goes wrong with the light bulb seven years from now.

The August 2005 issue of *National Geographic* magazine included a photograph of a man in front a huge pile of coal holding a compact fluorescent light bulb. The caption read, "A CF light bulb lasts 10 times longer and saves nearly a quarter ton of coal over its lifetime." I contend that everything I'm recommending in this book has three huge benefits. Replacing your light bulbs with CFLs is going to save you money. This helps you deal with the challenges we're facing economically. It also significantly reduces your carbon footprint. That's good for the planet. In this case it also helps you reduce the costs to operate your home. Good for your pocketbook. Good for the planet. Good for reducing your operating costs to help make you more independent.

As you read the energy-efficiency chapters of books like *$mart Power* or *The Renewable Energy Handbook* you discover numerous ways to increase your home's efficiency and save money. Some of these involve water-saving devices like low-flush toilets and low-flow showerheads. In the city the cost of water is generally so cheap that you will not notice a huge difference in your bill if you reduce your water consumption, but that will change as municipalities find it increasingly expensive to maintain aging infrastructure. In the country, water can be a much bigger issue. In our off-the-electricity grid home one of our largest uses of electricity is our deep-well pump, so using water efficiently is really important. If someone else supplies your water you should still be concerned about using water wisely because so much energy goes into getting it to your house. The city of Toronto, Ontario has a large transit system with subways, streetcars, and electric buses. It uses more electricity to pump water through the city's water mains than it does for everything else including the transit system, street lights, and powering its municipal buildings, including pools and arenas!

Basic energy efficiency also allows you to start evaluating just how committed you are to a path of independence. There are a number of products in your home that have a negative impact not only on the planet but also on your ability to be independent. They are machines that you have to keep pumping money into even though there are perfectly good alternatives that only cost you your time. Many see something like using a clothesline rather than a dryer as a step back in human progress but in reality many of us are going to have to relearn these skills anyway. The question is do you want to embrace it yourself or have it forced upon you?

Drying your clothes is an excellent example. There is no more wasteful activity in your home than drying your clothes in a dryer. You use fossil-fuel energy that you have to pay for and you super heat air for the sole purpose of quickly removing the moisture from the clothes you've just washed. This is an activity that will happen quite naturally on its own and much faster if you use the sun and wind to help. We take that super heated air and we simply pump it out to the atmosphere. It's a one-way trip to heat up the planet. And you've had to pay for the luxury. At least when you burn fossil fuels in your furnace the heat that it creates warms your home. But a dryer, that's just a waste of energy. And worst of all it causes your clothes to deteriorate much faster than they would otherwise. When you empty that "lint trap" on your dryer you're basically throwing away your clothes, piece by piece.

Now a clothesline, there's an elegant piece of technology. A simple way to accomplish a task naturally, letting the sun and wind do what they do naturally. And you don't even need the sun and wind for that matter. There are numerous drying racks that allow you to dry clothes in your house or apartment. You just drape the clothes over the rack and moisture escapes into the air. When you use these in the winter it's a nice way to add to the air moisture that your furnace or woodstove removes.

A dryer represents everything that's wrong with North American culture. We're all in too big a rush. We have to get everything done right away, now! Doing a laundry and then not being able to put the clothes away for eight hours is simply not on the agenda. And because we're in a hurry we're going to use massive amounts of energy and add to climate change while we're at it. And it's going to cost us. Oh yes it is. Not just the actual money we spend on the natural gas or electricity, but on the time that we need to make that money. Maybe, just maybe, if we worked a little less at a paying job, we'd have more time to hang that laundry out on the line. And if you are one of the millions of unfortunate North Americans who have lost their jobs in this economic downturn, you'd better be hanging your clothes on a line or drying racks. There is no excuse for spending a penny on a dryer.

And how about that dishwasher? Do you really need a dishwasher? Are you that busy that you can't take a few minutes to wash dishes by hand? And don't get confused by those silly reports that dishwashers are actually better for the environment. That's an absolutely ludicrous concept. Some European appliance manufacturer has been able to create a bogus study and one of the greatest urban myths of all time. Think about it. If you wash your dishes by hand you use hot water. If you use a dishwasher you use hot water, and lots of it, much of it super heated enough to blast off the macaroni and cheese that's been drying on that plate for two days. So the dishwasher is not only using energy to super heat that water, it's also blasting that water around. It's using electricity to power all that activity, as opposed to the activity that you create when you move the scrubbing pad over that plate.

Your home is actually full of appliances that do things you could do for much less money and with a much smaller carbon footprint. Ceiling fans use significantly less energy than an air conditioner. Using a push mower to cut your grass has a huge impact on reducing pollutants and CO_2 emissions. A rake will move leaves with much less damage to the planet than a leaf blower. Same with a shovel for snow removal rather than a snow blower. Start looking around your home and figuring out what you can do manually rather than using some machine. And you'll not only protect the health of your pocket book, you'll also protect your own personal health and the health of the planet. All the activities that we're talking about are much better for you. Sucking in the exhaust from a lawn mower or a leaf blower's internal combustion engine is not good for you.

If you ask a lot of people why they don't cut their lawn with an engine-free push mower you'll find it's because they don't have the time. They have to get the lawn cut quickly so that they can jump in the SUV and get to the health club. Well, if you want exercise doesn't it make more sense to ride your bike to the health club? And doesn't it therefore follow that if you want to burn calories you should cut the lawn with the push mower and ride your bike to pick up milk and a DVD rental? Not only have you saved on the health club membership, you just took a huge bite out of your carbon emissions from both the lawnmower and the vehicle. And the money you saved on the fitness membership went right into your "Solar Panel Fund."

Sometimes the simplest solutions are the best. Good for your health. Good for the health of the planet. Good for your pocketbook.

Once you've worked through the basic energy-efficiency savings it's time to look at advanced energy efficiency, which includes appliances. Appliances have a natural life and you have to replace them eventually. Tough economic times will force many people to hold on to them much longer than they otherwise would, but if your fridge is a beautiful 1970s avocado green, it's time to replace it. Over the long haul as the cost of electricity rises older models are going to be huge drains on your Ceiling fans.

When it comes time to replace your appliances remember that it's not the purchase price of a product you have to consider, it's the lifecycle cost of that product. What will that product cost to operate over its lifetime? If you buy the most efficient washing machine, with the money you'll save in operating that machine over its lifetime you could buy three of the least efficient machines. This assumes you are using hot water and have four people in the home. You'll also get a fivefold reduction in water use and a tenfold reduction in soap use.

The way to determine if the appliance is the most energy efficient is to use the EnerGuide rating that should be posted on or in the appliance. This will show you how this particular model stacks up against other comparable models in its class. You want to make sure you buy the model that uses the least energy. Sometimes you will not be able to buy the very best, but as long as you're close you're on the right path to keeping your ongoing energy costs down.

You probably won't be able to replace all of your appliances right away, but as they get towards the end of their life it's important to replace them with the most efficient models. You should also be looking for the EnerStar logo as an indicator of efficiency in the product class. This is helpful when you're looking at electronics like TVs and computers. Laptop computers can be up to five times more efficient than desktop computers. If you have to use a desktop make sure you look for an LCD screen similar to the kind laptops use rather than the traditional large CRT type.

If you want to get really hard-core about not wasting energy you should also deal with your phantom loads. North Americans are always in a big rush, so appliances like our televisions have an "instant-on" feature that keeps them ready to turn on the second we hit the power button on the remote. I guess we just don't have the time to wait for TVs to warm up anymore. With the instant-on feature, the television basically has to be left on constantly, using up to 80% of the power when turned off that it will use when turned on. You can prevent this either by putting it on a power bar which you turn off or by having the receptacle the TV plugs into wired to include a switch higher on the wall near it, which you turn off when finished. Up to 6% of the electricity generated in North America goes to these "phantoms loads," accomplishing nothing, adding to climate change, and draining money from your wallet. As you walk around your home at night and see all the flashing LEDs on VCRs and microwave ovens and see all the red LEDs glowing on cordless phones, rechargeable razors, and night lights, you're seeing energy being wasted.

I always recommend that you purchase power bars where the ON/OFF switch is lit to tell you whether it's turned on or off. That way as you leave a room you can easily check that the phantom loads are dealt with.

All these things add up to big energy savings in the long run and to your keeping more of your after-tax income. So why not quantify that savings to reward yourself for your accomplishments and inspire yourself to move onto the next step? I recommend that you open a brand new bank account and call it your "Green Energy Account" or your "Solar Powered Bank Account." Put all the money you've saved through your energy-efficiency activities towards products that will now generate green power and make you even more energy independent.

So before you start any of your energy-efficiency projects, make sure you keep the last few years' worth of energy bills. Then, as you begin reducing your electricity consumption, calculate how much money you are saving month to month. This can even extend beyond electricity consumption. If you have had an energy audit done on your home and you have upgraded insulation or improved weather stripping and replaced leaky windows, try to compare with previous years energy consumption and calculate the savings. Since natural gas prices in particular will change, look at how much of the energy source—gas, oil, or electricity—you have saved and then use the current, probably higher, price to calculate your savings. This is money that you've earned and paid tax on, so these savings represent real progress in your move towards independence on all fronts.

An energy audit is an excellent tool to help in your quest for efficiency. Many governments offer rebates on audits where a professional energy expert will come into your home and test it to see how efficient it is. This is often done with a door blower test where the house is pressurized to see how much air leaks out over time or how quickly the air turns over. Based on the test you will receive recommendations on how to improve the efficiency of your home. Some governments will give you rebates of a portion of the amount you spend to make the upgrades. Make sure before you start improving the efficiency of your home and adding renewable-energy equipment that you determine what incentives there may be from all levels of government. Some programs will be time-sensitive and will have to be completed within a certain amount of time after the energy audit. So if you plan on making some large expenditures on upgrading insulation or windows or adding a solar domestic hot water system, make sure you'll have the funds available within the allotted time.

As you go through the process of making your home more energy efficient, eventually you'll hit the point of "Maximum Energy Efficiency" where you've pretty much done all you can to save every kilowatt of electricity and every penny of wasted income. This has probably taken a year or two, especially with appliances, but you should now have accumulated a nice amount in your Green Energy Account. So if you go back to the chart on page 108, you'll see that you're now ready to start generating some of your own energy rather than just saving it.

Solar Domestic Hot Water

Most people assume that putting a solar panel on your roof is for generating electricity, but the first panel you put on your roof should actually be to make hot water. This is called "solar thermal" or solar domestic hot water (SDHW), meaning that it is for your daily hot water needs like washing clothes and dishes and bathing, not for heating your home. If you've ever picked up a hose that's been on the ground on a sunny day and felt how hot the water is when you first turn on the tap, you know about the sun's ability to heat water. Using this principal you will actually get a much faster payback when you invest your green energy dollar in a solar thermal system first.

The system consists of a collector that goes on the roof to capture the sun's thermal or heat energy. Then there's a system to transfer that heat into your hot water tank. Some systems pump the water through tanks with coils inside to heat the water; some use heat exchangers on the outside of the tank. There is also a choice of two main systems to go on your roof. One is a flat-panel collector and the other consists of vacuum tubes. Both have advantages and disadvantages, and some research about your local conditions and a discussion with local dealers will give you an idea of what is right for your home. The main systems on the market now are all 12-month units, meaning that they work year-round. These systems circulate a food-grade antifreeze called propylene glycol through the system and then transfer the heat to your domestic hot water supply.

Early solar thermal systems just ran the water for your home through a panel on the roof to heat it. Obviously it would not have been a good thing to have water freeze in pipes on your roof if the temperature got below freezing. So these systems were often referred to as "drainback," meaning that once the sun went down or on cold and cloudy days the water would drain back from where it might freeze and then be pumped back into the collector once it was safe.

Solar thermal systems are simply amazing products. There is nothing more gratifying than trying to hold onto a hot water pipe that is bringing sun-heated water into your house, with no fossil fuels burned and therefore no greenhouse gases flowing out of your chimney or greenbacks flowing out of your wallet. And best of all it increases your independence from rapidly rising energy costs and potentially disruptive shortages of the fuel source you use for hot water. The size of the system you should install will depend on the number of people living in your house and how extensively you use hot water. A house with four teenagers will use a lot more hot water than a home for an older couple. To a certain extent, some people may find that they start using hot water in tune with the weather. If you wait until a sunny day to wash your clothes, you'll have lots of hot water as well as heat from the sun to dry the laundry on the clothesline. This is a good thing.

The payback on these systems is impressive as well. Using Bill Kemp's chart you should be looking at about a six- to seven-year payback on your investment in a solar thermal system, assuming energy rates stay constant. I would argue that with peak oil and peak natural gas it's highly unlikely that energy rates will go anywhere but up. You may therefore find the payback is even faster. This assumes you've purchased a commercial unit and had it professionally installed. These units are not necessarily that simple to install, and you should have a fairly good working knowledge of plumbing if you decide to do it yourself. You should also be comfortable working on your roof. If you are, you can save a lot of money by installing your own system.

Some people may even attempt to build their own system using off-the-shelf components. These units are usually less sophisticated and are drainback systems to avoid having to charge the system with propylene glycol, but that's not to say if you're up for the project you shouldn't do it. If you've lost your job and are having trouble finding work but your spouse is employed, you'll be no doubt cutting back on expenses everywhere. This may be a case where you invest your time and some money which over the long haul offer you an excellent payback. You're just taking the time you could otherwise spend earning an income to pay someone else to install your system and using it instead to learn how to do the installation, allowing extra time to make some mistakes along the way.

Having a shower is a wonderful thing. Standing there as hot jets of hot water cascade down upon you with the smell of cleansing soaps and shampoos is a huge luxury. The problem is that most of us don't see showers as a luxury. We forget how many of our fellow human beings don't have access to this most basic personal hygiene. But since we're embroiled in the tensions of a hard economic climate, it's always nice to be able to relax and de-stress in a shower. So having a shower should be a priority for you in these trying times. Having one that is pretty much independent of outside suppliers is even better, and a solar thermal system provides that. It saves you from purchasing electricity or natural gas or propane to heat your hot water. It requires a huge amount of energy and therefore expense to heat hot water in your home. You are probably spending at least 15% of your energy budget on hot water, maybe more, so the investment in solar thermal is going to significantly decrease your monthly expenses once the system is paid off.

Standing under a shower where hot water is being heated by the sun is an even better thing. You are not releasing any emissions of CO_2 into the atmosphere, and the shower isn't costing you a penny. Investing in a solar thermal system is one of the soundest financial commitments you'll make. Your retirement fund statement showing you how much your stocks have gone down may make you red in the face, but it won't be as comforting as a hot shower on a cold, sunny day.

On Bill Kemp's chart you'll also notice that wood stoves and ground-source heat pumps also appear in the "thermal" section. Both of these make use of solar energy and they are discussed in more detail in the heating section in Chapter 11.

Solar & Wind Electric

The second type of "ECO-nomic" payback afforded by renewable energy comes from using the sun and wind and water to produce electricity. While you may invest anywhere from $3,000 to $7,000 in a solar thermal system, you're probably going to want to start at that $7,000 figure and work up with this equipment. A figure like $20,000 or $30,000 is more in line with what many invest in these systems. There is still a payback, but it takes a little longer. The comfort and security this will bring to your home is priceless.

Solar or photovoltaic (PV) panels are the best place for most people to start using renewable energy to generate electricity. A solar panel converts the sun's energy into DC (direct current) electricity, but most of the appliances in your home require AC (alternating current). You use an inverter to convert the DC to AC so that you can power all the AC appliances in your home. In the early stages of the industry some people used the DC electricity directly and purchased DC appliances, which were commonly used in the recreational vehicle industry at the time. Early inverters were quite crude and inefficient, so early on this made sense, but modern inverters are extremely efficient and there is minimal loss involved in making the conversion, so it only makes sense to go AC.

Many levels of government now provide financial incentives for the installation of renewable energy, so you need to research what is available where you live before you proceed. Some incentives provide a rebate on a portion of the purchase price of equipment, while others actually purchase your "green" power for a specified price over a given contract length. Your dealer will be able to help you calculate the savings this will represent. The Database of State Incentives for Renewables and Efficiency (www. dsireusa.org) is an excellent resource to find incentives in your state. In Canada The Office of Energy Efficiency of Natural Resources Canada provides a great listing of incentives (www.oee.nrcan.gc.ca.)

As outlined in *The Renewable Energy Handbook*, you'll need to calculate your electrical loads before a dealer can properly size your system. This is where taking all that time to reduce your electrical usage will pay off. For every dollar you spend on energy efficiency, you'll have to spend about $5 for generation. In other words, you can continue to use inefficient incandescent light bulbs, but you'll need to spend $5 more to purchase solar panels versus the $1 you spend on energy efficiency to reduce your demand.

Once you've reduced your electrical requirements you can make your investment in solar panels. It is an investment. In fact, when you look at what's happening in the financial markets these days, the payback is outstanding. And best of all, they will be providing a huge degree of energy security for your family. Anyone who has experienced a blackout knows how quickly the inconvenience of being without power turns into a huge problem. Solar panels are hard assets, and hard assets with numerous advantages. When it comes time to size your system I recommend you oversize it. Having more power than you need is not a problem and you can always find productive things to do with it, whether it's pumping it back into the grid or preheating hot water.

One of the biggest concerns I hear from people contemplating investing in PV panels is technology. They hear about thin-film panels and nanotechnology and all sorts of other fancy technologies driving down the price and increasing the efficiency of solar panels. And no doubt this will be true to a certain extent; but I would suggest that the enhancements will be gradual and do not warrant any hesitancy on your part. Yes, panels will get more efficient, but it will be incremental. For many years the assumption was that the price of panels would come down dramatically, but that hasn't been the case. With progressive governments in Germany and Japan and California providing excellent incentives, the worldwide demand for solar panels has remained high, to the point of there being a shortage of the silicon that is the key ingredient in their manufacture. Economic dislocation such as we're experiencing now often causes a reduction in demand which deflates prices, so now may be a great time to purchase PV.

You should not wait to take the plunge. Think about computers. Moore's Law predicted in the 1960s that computer processing power would double every 18 months, and amazingly that has continued. Based on Moore's Law, you should never purchase a computer because within 18 months the newer ones will be twice as fast. The logic of this would have you typing all your correspondence and mailing it to your friends and family as well as trekking off to the library to research things in books. Instead, at some point you no doubt jumped in and learned how to use this amazing tool and upgraded it as the technology advanced.

So don't wait. Solar panels work! They make electricity... today! They help inflation-proof your family, today. They reduce your carbon footprint today. They'll power your home to keep the lights on next time an ice storm takes down power lines or a hurricane causes outages. They are an indispensable tool for your home, today. They are the ultimate "hard asset" for challenging times.

You can mount solar panels either on your roof or on a tracker. A tracker allows your solar panels to follow the path of the sun, but it requires a fairly wide unobstructed path, so you need a fairly large lot to use one. It is also additional expense and one more thing to break. What's generally more convenient is to mount solar panels on your roof. They need to face south and should not be obstructed by trees (or nearby apartment buildings) to get the maximum return on your investment. Some people who don't have a south-facing roof actually use solar panels like awnings, mounting them over the windows on the south side of the house. This is an excellent way to reduce your air conditioning loads, keeping the sun out of those windows in the summer when the sun is high. In the winter, when the sun is lower, it should be able to beam into your home under the panel awnings and warm it up. This is the basic principle of passive solar design in new homes.

Along with your solar panels you'll need a few other pieces of equipment like a charge controller and a DC disconnect to isolate the solar panels from the rest of the electrical system. While it's possible to not have batteries connected to your solar panels, I would recommend you have some. If you're connected to the grid you won't need as many as if you're off grid, but you should have some nonetheless. Someone who lives off the grid and is not connected to the electrical utility will need a larger capacity for their batteries to allow them to run their households for several days of cloudy weather. They might spend $5,000 or $6,000 on their battery bank. Someone who is connected to the grid should use the grid as their large battery storage, putting excess electricity into it during the day and taking out what they need at night. They should also have a small battery bank, perhaps under $2,000, to allow their household to run should the grid go down but not have more storage capacity than they need.

If you're grid-connected you should also get an electrician in to rewire your panel box to have "essential" and "non-essential" loads. This will ensure that your batteries will only have to power loads that are absolutely critical for you during a power outage. These would include your fridge and freezer, a furnace fan if you heat with a forced-air furnace, lights, and perhaps a TV or radio so you can stay up-to-date on what is happening to get the power back on. Sizing the battery bank to power these loads will depend on how efficient they are, so you should be powering a very efficient refrigerator and light bulbs to make sure you aren't wasting any of the limited power you've got in your batteries. It will also depend on what type of outages you expect. If you're used to summer storms leaving you without electricity for a few hours it will be quite a bit different than if you experience hurricanes or ice storms leaving you in the dark for days.

The nice thing about having essential and non-essential panel boxes is that it will make it easier if you decide to add a generator to the mix. Lots of books by "doomers" who anticipate a rapid breakdown in social order suggest purchasing generators. This sounds good on the surface, but you need fossil fuel to run a generator and where will you get that? A better plan is to have a reasonably sized generator and renewable energy that will run your household loads and charge your batteries. That way if the power outage lasts for a few cloudy days and your solar panels aren't charging your batteries you'll have a backup system to keep your house humming along. People who experienced the ice storm of 1998 that crippled much of the Northern U.S. and Eastern Canada spent many days in the dark, and in some cases even weeks. The freezing rain lasted for days, so if you did have solar panels a generator was key to operating your household.

That ice storm inconvenienced millions of North Americans and left hundreds of thousands in dire straits. Rural people without electricity couldn't pump water. People who heated with natural gas or oil furnaces had no heat. Food went bad in freezers. People couldn't cook, couldn't bathe, and couldn't get around, because you need electricity to pump gas at gas stations and they didn't have power either. Eventually the military was called out in Canada to go door to door to help those in rough shape.

A few generations ago people were independent. In the rural community where I live people would have had a well to supply their water. They would have heated with wood they cut on their property. They would have had a pantry with food basics like flour and sugar and fruits and vegetables that they had preserved. They would have had a root cellar. They would have had potatoes and carrots and root vegetables that lasted right through until spring, and many would have had animals to provide milk and eggs and meat when they needed them. In other words, if there had been a bad ice storm they might have spilt some milk on the slippery walk back from the barn, but they would have been largely unaffected.

Today, on the other hand, people have become completely dependent on other people for most of the essentials in their lives. Heat, water, food—these are all things we now rely on others for. The theme of this book is that you have to get these back under your own control because you will not be able to rely on those outside providers to continue in the way you've become accustomed to. You need to think the way your grandparents did. You need to make your grandparents proud.

I like to tell the story of our friends Bill and Lorraine Kemp, who live off the grid near Ottawa, Ontario and were in the heart of the ice storm of January 1998. The military was having to evacuate people to shelters because they were freezing in their homes and had nothing to eat. Houses with no heat eventually had water pipes freeze and burst. It was not a pretty scene. When the military arrived at Bill and Lorraine's house Bill went to the door in his bathrobe because he had just had a shower. Heat from their woodstove wafted from the house, Mozart blared on their stereo and Bill had his typical steaming cappuccino in hand, something he is rarely without. The two soldiers looked at their clipboard, looked at Bill, heard the music, felt the heat, smelled the coffee and announced, "Well, we think we know the answer, but is everyone in the house alright?"

"Never better!" was Bill's reply, and this should be your mantra when you've finished reading this book. When someone arrives at your home to see if you're all right, make sure this is your response. Better yet, you should be the one inquiring of your neighbors to ensure that they're weathering the storm. This is what neighbors do, and those solar panels and that battery backup will give you the luxury of being the one offering the help rather than receiving it.

You'll also have to give some thought to how you power that generator. It will be either gas, diesel, natural gas, or propane, but it will be a non-renewable fossil fuel. The bulk of smaller home-sized generators will run on gas or diesel, so you should have a number of jerry cans of fuel on hand. The challenge is that neither of these fuels will last forever. Over time the fuel will become unstable and will not work in the generator or your car's engine. So if the fuel is going to sit for a while, add fuel stabilizer to it. This should allow you to store it for one to two years. I would suggest you not store it that long, and cycle it through other gas engines you might have like your car or lawn mower if you're still using one. Every three to six months use the stored gas in your car and refill the storage containers. Try and store the gas in as cool a location as you can find, out of the sun. If you have a diesel generator make sure you purchase the yellow plastic containers to differentiate it from gas. Put a tag on the jerry can that indicates when you purchased the fuel and whether or not you've added fuel stabilizer to it. And remember, you have to be careful where you store a flammable product like this. Do not store it anywhere near a source of heat or ignition. Never fill your generator in a garage or enclosed area because of fumes or in case of a spill. Don't smoke near these containers.

The key to a generator working well is to run it periodically. Too many people have generators that sit in a garage for months, and when the power goes off they can't get them started. With any internal combustion engine you need to use it periodically to keep it in working condition. If it is an electric start you'll have to make sure the battery stays charged as well. You should have a battery charger that you can use to keep it charged. You should also have jumper cables so that you can jump start the generator from another battery if you can't get it started when you need it. Electric starts are an excellent option on a generator because pull starts tend to be fairly difficult, or at least require a pretty good yank. So if you have back problems or a spouse who doesn't want to be stuck whaling away on the pull start while you're away at work, try and get an electric start, and keep that battery charged and ready to go. If it's a smaller electric start don't assume that the generator will charge the battery when it's running. Keep your eye on the battery voltage to make sure it's high enough. Start that generator every two or three months to make sure it's ready when you need it.

Another excellent way to keep those backup batteries charged and help with your energy independence is a wind turbine. I should qualify this by saying that wind is excellent if you are in the right location. It's the dream of many people to have a wind turbine at their home, but if you live in an urban area this may not be realistic. The best wind is the wind that has had a long open area to travel over before it hits a turbine. A large body of open water is the best location. If you live beside a large lake or river you are probably in a good spot for wind. If you live in an area with lots of open fields wind is something to consider. Wind also accelerates up a hill, so if your home is perched on top of a hill consider an investment in wind. What wind doesn't like are obstructions like forests and barns and houses. Wind flows like water in a stream. When water encounters obstructions like rocks it slows down. The rocks create turbulence and reduce the flow rate, known as "laminar flow." Wind operates the same way; every time it encounters an obstruction turbulence is created, which interferes with its flow.

Urban areas are composed of endless obstructions that create turbulence, so they are generally not a good area for wind. Many cities are located on large bodies of water, but the areas around the water tend to be the most economically desirable, so putting wind turbines there poses a problem. If you work in a large building downtown you know how windy it can be around those office towers, but this is not good wind for power production because it is so erratic. All that turbulence would confuse a wind turbine that wants to have a strong consistent wind turning its blades from one direction.

Most wind turbines today are the well-known three-blade horizontal-axis turbines like the very large ones you see in wind farms that are becoming more common in many parts of North America. They represent the end result of years of experimentation and refinement. Many designs have been attempted, but large turbine manufacturers have settled on this design because it has proven to be the most effective way to generate electricity from the wind. Most smaller home-sized wind turbines follow this design as well.

There are a number of new "vertical-axis" turbines entering the market that are more of an egg-beater design, with the blades rotating around a center point that goes up and down, rather than left to right. These are touted as home units that can produce power in very little wind. The challenge with this claim is twofold. Most of these turbines would be mounted on homes in urban areas where the wind resources are marginal. And while it might look great to have a turbine spinning on your home, even slowly, there is basically no power in that lighter wind so the electricity the turbine will be producing is marginal.

As is so often the case when you have a sudden interest in something like the environment, the market is flooded with products that seem to help solve the problem but have unproven technology. If you are looking to purchase a wind turbine for your home you'll want to see its "power curve," which will show you how much power it's capable of producing over increasing wind speeds. Most turbines will ramp up to a peak output and then drop off as the turbine reaches its maximum-rated speed, stalling or furling to prevent it from damaging itself in excessive winds. Some of the newer vertical-axis turbines have been slow to produce power curve data, which means that you should be skeptical about their rated performance. The home-turbine market is dominated by several manufacturers that have been producing smaller-scale turbines for many years, including Bergey Wind, Southwest Windpower, Kestrel, and Africa Wind Power. If you are considering a wind turbine I would recommend that you stick with one of the main manufacturers with a proven record. These companies all produce turbines that use the standard horizontal-axis design.

The correct way to determine if your site has wind resources that make the investment in a wind turbine worthwhile is to put up an anemometer at the height you anticipate installing the turbine and logging the wind speed over time. This can be cumbersome, and often people go to online resources that include wind maps of your area. Various levels of government have maps which allow you to decide how good your local wind resources are. Many people say "the wind always blows at my house," but whether or not that wind has good potential energy in it remains to be seen. A wind turbine should be 30 feet above and 300 feet away from the nearest obstruction such as a barn or silo or a forest to ensure a good clean flow of air to maximize the potential of your turbine investment.

Several years ago we put up a new Bergey XL1 (1 kilowatt) wind turbine. We had a smaller turbine on a 60-foot tower that never lived up to its potential. It broke twice, and the second time I decided it was time to go big or go home. We installed the Bergey on a 100-foot tilt-up tower. We are not in a prime location for wind. In fact I believe we would be considered a poor area. We are 45 minutes north of the Great Lakes, which are an excellent wind source, but we are too far away to take advantage of them. Forests, thousands of acres of forests, also surround us. While these make for an amazing place to live, they are not conducive to wind power. Our tree line is about 60 feet, so we felt that with the tower 30 or 40 feet above them we at least had a shot at some wind.

Living off the electricity grid we also have a unique perspective on wind. In any off-grid home having a hybrid system that utilizes both solar power and wind power makes sense. The months that have the most wind, in our case the fall and winter, also have the least amount of sun and vice versa. Before installing the new turbine we needed to run our gasoline generator about 15 times during the fall months in order to charge up our batteries during periods of cloudy weather. After installing the turbine in September 2007 we ran the generator only three times, and then went ten months without running it at all. This was quite an accomplishment for us. Our desire to reduce our generator run time comes not only from a desire to add less carbon to the atmosphere but also from a fear of where we see the price and availability of oil going. With most oil-producing countries past peak and many in steep decline, I believe I eventually won't be able to afford to run the generator, if I can find gas at all. So our wind and solar ensure that our lights stay on, our pump keeps water flowing in the house, and the fridge and freezer preserve our food.

Installing the wind turbine was a huge challenge because although I am not an engineer I decided to install the turbine myself anyway. Many local dealers are at the stage where they don't want to be bothered with wind. Wind turbines are mechanical instruments, which means they break. And when they break they are usually located at the top of poles and towers that make them difficult to reach. With solar panels, on the other hand, once you mount them on a roof or tracker they just work. This might be something you should bear in mind as you evaluate various technologies to make your home more independent.

After living off the grid for ten years I had developed a certain amount of confidence that I was up to the task and knew it would save a lot of money if I installed the turbine myself. With our video publishing business, we also decided it would make an excellent opportunity to produce a new DVD called *Installing a Home-Scale Wind Turbine*, which we added to our catalog. While putting up a wind turbine is a lot of work, videotaping the whole process doubles (or triples) the work because you have to move the equipment into position, bring out the tripod, set up the camera, set up the shot, and half the time do the step you're working on more than once to a get a good take.

We had to use a backhoe early in the process to level out the area where the tower was going since it helps if all four anchors are roughly level. Then we had to dig six holes, including one for the base and one for the winch. I elected to dig these holes by hand because we were experiencing a drought and our sandy soil was collapsing into the hole when we had the backhoe attempt it. I learned that digging a 5-foot hole to get below the frost line is easy in sand for about the first 2½ feet, but when you're only 5'8" you can start getting tired rotator cuffs pretty quickly when tossing sand over your head. We built rebar cages to strengthen the concrete we poured into the holes.

Once the anchors were in I had to dig the trench from the base of the tower to the battery room, and at this point I decided it was time to get the backhoe back. Then the tower had to be laid out and guy wires prepared for the lift. We lifted our turbine with a permanently installed winch which we power with an 18V Dewalt cordless drill. We can raise the tower with four batteries and lower it with two, so it's an exceptionally efficient winch. The nice thing about our gin-pole-type tower is that we can raise and lower it fairly conveniently. This helps if you ever have to do maintenance on the turbine or if you are in a hurricane-prone area. You could have the tower down in about half an hour if a huge windstorm approached.

When the tower was finally up I must say it was one of the most gratifying things I had ever done. For years we had been exhibiting our books at renewable-energy fairs and usually took solar panels and our old wind turbine there to attract attention to our booth. This is where I learned the male/female side of renewable energy. If you have a working wind turbine at a show, men will be drawn to it like moths to a light bulb. They will spin it as hard as they can, stick their hands into it to see how much it hurts when they try and stop it, and try and figure out how to make their own turbine from the broken car alternator in their garage. Wind turbines are very male. They can be loud, make a lot of commotion, and scream "look at me!" Women, on the other hand, are drawn to solar panels, which work away quietly, generating electricity calmly and efficiently. They just do the job. They don't make a lot of noise, don't draw attention to themselves, just work away quietly in the background.

With two daughters, I have always considered myself a feminist and this certainly held true with my solar panels. I love my solar panels and marvel at how quiet and efficient they are.

But my wind turbine…. I really, REALLY love my wind turbine. I am a huge disappointment to feminists everywhere because of how much I love my wind turbine. When it's whirring away in a high wind I can watch it for hours. I can go over and stand at the base and gaze up at it. I can put my hands on the tower and feel the energy coursing through the steel and being carried safely by the wires inside into my battery room, where it can accomplish great tasks in the house. We have an AMP meter in our battery room which records how much power the turbine is producing. Our winds are often not consistent, sending the meter up and down constantly, with me cheering when it hits the high end of its capability. In fact, even with a satellite television dish and 100 channels, there is nothing better to watch on a windy night than that AMP meter dizzily dancing up and down.

I share my enthusiasm for my solar panels and wind turbine with you to emphasize a point. For many of us our jobs completely remove us from the possibility of any great sense of accomplishment. Filling out spreadsheets, filing sales reports, pushing bits of information around a computer endlessly can be disillusioning and demoralizing. For what? What is the end result of the investment of my time? Yes, I am earning an income, but if I'm robbing my soul to accomplish it then I'm not going to be very enthused about life. This is an opportunity renewable energy offers you. It gives you a chance to invest some time in learning about it and then integrating it into your life. That will pay you back in more ways than just a good return on your investment. It will fill your soul with joy as you watch your carbon footprint shrink. It will fill you with confidence that you are making your home and family more independent of outside suppliers of energy who may not be able to provide a consistent product in the future. It will bring joy into your life in a huge variety of ways. But you need to get started, right away.

The first step you should take is to make sure you have some battery power packs around to run your really essential loads. These will include both a battery and an inverter that you can use to run small appliances and charge things like cell phones during power outages. Then you should develop a plan to begin building your renewable energy system. If you have limited funds but want to get started on a system that will be expandable, start with a good inverter. By this I don't mean the kind you can get at a big box store for $200; I mean the kind you'll get through a renewable-energy dealer for $2,000. This more expensive inverter will have features that make it far more valuable than just taking the DC power you have stored in your batteries or that the solar panels make and converting it to high-quality AC. It will have features like "surge capacity." While an inverter may be rated at 2,500W (enough to run your washing machine and microwave), periodically your home will have very large loads. For someone living in the country, for instance, when the pump comes on there will be a huge surge of power required to get the pump moving water. Once the water is flowing it consumes much less power over time. A good 2,500W inverter will have a 6,000W surge capacity to handle those large loads. It may be just for a few minutes, but it will allow your home to function properly when those surges happen. A cheaper inverter is more likely to shut down during a large surge because the load exceeds its capacity.

A good inverter will also be able to charge batteries through your generator. This will ensure that your batteries get the proper charge. If you are connected to the electricity grid a good inverter will be able to handle all the work that must be done on a daily basis. Once the sun comes out it will first make sure your batteries get charged, then once they are charged it will run all your household loads, and if there is more power than your home is using it will divert the excess to the grid for your neighbors to use. On a cloudy day when you turn on the washing machine the inverter will automatically know there is not enough power coming from the solar panels so it will go to the grid to get the electricity you need. A less expensive inverter does not have this capability.

If you invest in a good inverter it will also be expandable. Let's say you spend $2,500 on a good inverter and $1,000 on batteries and $1,000 on solar panels. This will keep the lights on and fridge running during a power outage, and the batteries will be charged when the sun comes out. Next year, when you get your $1,000 income tax refund, you can add more solar panels. The following year you can take that money you set aside in your "Green Energy Bank Account" and buy more batteries to increase how long you can ride out a power outage. Then a few years down the road when your uncle Harry dies, you can take the $5,000 he left you and put a whole bunch more PV on your roof. The beauty of investing in a good inverter is that it is expandable and will allow your system to grow without having to be upgraded.

An excellent resource for learning more about renewable energy is

The Renewable Energy Handbook by William Kemp. I will admit my bias since I publish this book. I will also suggest that it's the best-selling book on the topic in North America because Bill has done such an exceptional job of putting together a really approachable and digestible book. He's one of those rare engineers who really understands all the issues with the technologies but can explain them in a way that makes sense to the average individual. I've been most fortunate to have been mentored by Bill for many years in the ways of green power.

The energy your home uses for cooking poses a unique challenge in the future. As we learned in the peak oil chapter, natural gas, which many of us cook with, is getting scarce and will become increasingly expensive. Propane will be the same. Electricity will definitely go up in price as well—the question is how much? I would suggest that governments will try and limit electricity increases somewhat because it is such an essential part of our lives and is especially important for lower-income people. So while cooking with an electric oven will get more expensive, its increases may be less dramatic than with fossil fuels.

The one thing you learn when living with renewable energy is that it takes a huge amount of effort to make "heat," or "thermal" energy. Many people who live off the electricity grid sort of cheat and live more "on propane" than "off the grid." They shift all their largest energy loads, those being thermal loads, to propane, which is liquid natural gas that can be delivered by truck and stored in tanks. Since many off-gridders are far from natural gas hookups, this is convenient. But shifting all your major loads to propane defeats the purpose, especially if you moved off the grid to reduce your carbon footprint. If a lot of your grid-supplied electricity comes from a source like wind power, hydro, or even nuclear, you actually would have a lower carbon footprint using grid power for your thermal loads than propane.

Propane is produced when a barrel of crude oil is 'cracked' into its constituent parts, and since it has to be delivered by truck often from sources far from urban areas, the reality is that it is going to become one of the least reliable energy sources. So trying to find an alternative is important. In our off-grid home we are in the process of trying to eliminate propane. We installed a solar thermal system to heat our domestic hot water. Next year I will install a water loop through my woodstove to heat my water as well. This will help on those days when there is no sun. In November and December we often have cloudy days but it's cold enough that the wood stove is going, and it will heat our domestic hot water at the same time.

We have also added more solar panels to produce more electricity. We do some of our cooking with our convection oven, our microwave, and our induction burner stovetop. We also do some of our stovetop cooking on the wood stove and have acquired a wood stove oven that sits on top of the wood stove and can be used for baking. We also have a solar oven, which I strongly recommend everyone have. You can find plans for them on the Internet. After several attempts at a homemade one we purchased one from Sun Ovens (www.sunoven.com). It is a fantastic design and can boil water in 75 minutes when it's below zero outside in March!

We still have the challenge of our propane cookstove. Some people going off grid and off propane use a wood-burning cookstove, which we will move to eventually. With 150 acres of forests, we have all the wood we could ever use. It's renewable, and since it's just releasing carbon it captured during growth, it's carbon neutral. Nothing sounds nicer than making breakfast on a winter morning on a wood-burning stove. Nothing sounds worse than doing it in August in a heat wave, but when that heat wave hits we'll be getting lots of sun so we'll have ample electricity to use all our electric cooking tools like the convection oven and induction stovetop.

As you saw from Bill Kemp's "Renewable Energy Pay Back Time" chart earlier, when you finally do get around to using solar and wind power for electricity the payback is longer, but there is still a payback. Depending on where you live, what sort of government incentives are available, and what you pay for electricity, the payback time might be anywhere from 12 to 20 years or more. Let's say there were generous incentives in your area and you made an investment of $20,000 with an expected payback of 15 years. Even if it were 20 years, it's still an excellent payback. Most solar panels you purchase today will come with a 25-year warranty, which indicates that the manufacturer has real confidence that the panel is very robust and will still be functioning long into the future. The original solar panels shot into space on satellites in the 1960s are still working, so there's every indication the panels will be producing electricity for you for decades.

Your investment in solar and wind does have a payback. Now let's look at other investments or purchases you make to calculate their payback. What is the payback on your riding lawn mower? What was the payback on your trip south last winter? What was the payback on your $10,000 granite countertop? And what was the payback on your last vehicle?

Many people will happily spend $20,000 or $30,000 on a new car or truck knowing full well that in 12 to 15 years it will have no value. There is simply no payback. But for some reason we expect an investment in renewable energy to have a payback. Well it does and it's excellent!

After you hit that hypothetical payback point of 20 years, you get "Free Power for Life." Over time the efficiency of the panels may decline a bit, but they will keep producing clean, green, emissions-free electricity for your house for decades after they're paid off. Once you get those panels installed you are inflation-proofing your family from rapidly escalating energy costs. Like so much of our society's infrastructure, electrical utilities have not been investing enough to maintain their power grids and generation capacity, and the way they'll deal with this is to pass these costs onto ratepayers. By generating your own electricity you are taking control over something that brings great comfort and convenience to your family. In uncertain times, it will be nice to have lights on, a refrigerator preserving your food, and furnace fans running to keep your home warm.

And what about climate change? Fifty percent of the electricity generated in America comes from coal, which is basically pure carbon that, when burned in a power plant, releases carbon dioxide into the atmosphere, creating climate change. When you use power from the grid you are enlarging your carbon footprint. When you power your home with renewable energy you are reducing it, dramatically. What value do you put on doing the right thing for the planet?

With the decades your solar panels will produce energy they will become something that gets passed down to future generations in the family. What price do you put on the creation of a legacy? So many Americans want to leave a legacy for their children and grandchildren, from family cottages to college educations. How about leaving a family legacy of clean, green, carbon-free energy for future generations? What price can you put on that investment?

I think what has to change is our perception of the true cost of electricity and our responsibility for its impact on the planet. We assume that because it was always cheap and someone else's responsibility we have no obligation to take control of it for ourselves. It's time to change that perception. Stop assuming that someone else will responsibly and reliably provide power to your home. Start creating it yourself. You can start small and work your way up, but you should regard the generation of electricity as your job. You can do it efficiently and cleanly. And it will make your long-dead ancestors proud because you are returning to a time when people had control over their own destiny and hadn't turned over their well-being to an economic entity whose first obligation is to its shareholders, not your family.

It's easy; you just have to make the commitment to take back this control. And it's gratifying. Think of the most gratifying things you've done—your first job, raising your family, coaching a sports team—nothing will be more gratifying than looking up at your roof and seeing those solar panels powering your home, reducing your carbon footprint, increasing your energy inflation protection, and making your family more independent to weather the next storm. Make your grandparents proud! Take back the power. Make it yourself.

13 Food

*I'm at the age where food has taken the place of sex in my life. In
fact, I've just had a mirror put over my kitchen table.*
Rodney Dangerfield

Been sipping tea all day from a pot steeping slowly on the stove.
Sarah Harmer, "Good Fortune", Cold Snap

According to the U.S. Dept of Health and Human Services a moderately active adult needs about 2,000 calories a day to live healthily.
That's pretty easy these days. In fact, it's so easy that a lot of us consume
substantially more than 2,000 a day. So how and what you eat is going
to have a great impact on how you weather the changing world we live
in. The one defining theme should be for you to "eat lower on the food
chain."

When you think of the food pyramid that health agencies have
published and revised over the years, at the top you have fats, oils, and
sweets, which should be used "sparingly." For someone like me with a
sweet tooth, this is enough to make me throw the whole concept out.
But as you progress down the chart you find food groups you should eat
more of. It recommends 2 to 3 servings of the dairy group (milk, yogurt,
and cheese) and 2 to 3 servings of the "protein group," which includes
meat, poultry, fish, dry beans, eggs, and nuts.

Then you should have 3 to 5 servings of vegetables and 2 to 4 servings
of fruit. The final group is the one I'm going to focus on, which is the
bread, cereal, rice, and pasta group. The chart recommends 6–11 servings
per day. This is the carbohydrate group, one that sends shivers down the
spines of so many people still living on the mythical Atkins diet. You don't

hear much about that Atkins diet anymore. For a while the media was obsessed with it. Manufacturers were falling all over themselves to come out with "low- or no-carb" products. The funny thing about fad diets is that in the long term they don't usually work. People fall back to their old ways of eating and they often crave the foods they should be eating. Certainly in the case of the Atkins Diet, people craved carbohydrates. And why do you think that is?

Could it be that humans are basically herbivores and not meant to eat animal products? While lots of people like to think we're omnivores, meant to eat both plant and animal products, a closer look at our physiology helps clear out the fog on that issue. The first argument people usually make in suggesting that we're carnivores is our "incisor" teeth, the big ones that protrude like Dracula's. Over time these teeth have become less and less pronounced, to the point where if you pick up People magazine and look at the most beautiful people on the planet, they all have teeth like cows. Cows have very straight, flat teeth, designed to eat plants. The teeth you see on humans today look way more like a cow's than a wolf's or a dog's. Wolves and dogs have sharp, pointy, jagged teeth, designed to rip away at flesh. Grab a mirror. Take a look. Do your teeth look as if they should be chewing bread and rice or ripping apart raw flesh?

The most telling point in the argument that humans were never meant to eat meat in large quantities is our intestines. True meat eaters like dogs and wolves and cats have very short, straight, flat intestines. It's almost as if nature has said, "You know, you don't want this saturated fat and rotting flesh sitting in your intestine too long, so let's get it out of there fast." Our intestines, on the other hand, are very long, and instead of being straight and flat they're twisted and puckered. All those little indentations are good places for nasty things to sit and do bad things. The human intestine is about 6 to 8 yards long (6-7 meters). Stuff is going to be in there for a while and nature would prefer it was plantbased rather than animal based.

But I am not here to debate whether or not humans are herbivores or carnivores or omnivores. I'm not disputing that we "can" eat animal products. We've been doing it since we stood upright. The question today is "should" we be eating animal products, or how much of our diet should consist of animal products? I'm recommending that you look at reducing your consumption of animal products for a variety of reasons. The main one is that it will be much easier to feed yourself in the future if you base your diet on plant products. I'll also discuss the health benefits of a plant-based diet in Chapter 17.

The food pyramid is a government construct that is based on consultation with "stake holders", not only health professionals but also the people who produce the food. It is the result of negotiation with groups that make a living selling what they grow, and the dairy and meat industries have very strong lobbies. In fact, prior to the current food pyramid there was a version that recommended smaller portions of animal products, but industry pressure forced it to be revised. It has reverted back closer to the version I grew up in the 1960s where every classroom had a chart on the board showing the food groups. The chart was essentially an advertisement for the dairy and meat industries, giving them equal weighting with fruits and vegetables and breads and cereals. We know now that if you eat that way your health may suffer.

The reality is that you don't need animal products. Vegetarians are extremely healthy. In fact as a group they are much healthier than meat eaters. According to the Framington Heart Study, the largest study ever conducted on the health of Americans, vegetarians live years longer than meat eaters. The American Cancer Society suggests that vegetarians have lower rates of many cancers. There are all sorts of well-known celebrities who are vegetarian, like Paul McCartney, Pamela Anderson, Forest Whitaker, Ed Begley Jr., and Shania Twain.

Many successful athletes are vegetarian. In fact the only man to ever win the Hawaiian Iron Man Triathlon six times is Dave Scott, who was a vegan at the time. Other well known vegetarian athletes include Martina Navratilova, Hank Aaron, Billie Jean King, and Joe Namath

Many athletes have learned to prepare for major events like triathlons by "carbo-packing." They sit down and eat huge amounts of rice and pasta and carbohydrates that their bodies will need to convert to energy during the race. So if a triathlete can run 26 miles on a plate of pasta, why do you think you need to eat meat to sustain your daily pace? In fact you don't, and a large percentage of the world's population until recently have not eaten much animal protein. They have structured their diet around carbohydrates and starches like rice and pasta and bread and potatoes. These were the mainstay of their diet, not meat. Meat was the exception to the rule, more of a special-occasion addition. This should be the way you start to move your diet, toward one based on plant foods rather than animal products.

The reason this is going to be so important in the future is that is takes a lot more energy to produce a pound of animal protein than a pound of plant protein. For example, according to the United States Department of Agriculture, it takes 7 pounds of corn to produce 1 pound of beef, 6.5 pounds of corn to produce 1 pound of pork, and 2.6 pounds of corn to produce 1 pound of chicken (www.ers.usda.gov/AmberWaves/February08/Features/CornPrices.htm). So since we know humans can live quite well as vegetarians, in a resource-constrained world wouldn't it make more sense to feed that grain directly to humans rather than cycle it through animals to produce food that may not be healthy for humans to eat? This certainly holds true on a global scale, but it's even more important on a personal scale. If you were to turn that yard of yours into a big garden, it simply wouldn't make sense to grow nothing but wheat which you then fed to a cow which would produce protein on a 12:1 ratio, 12 pounds of grain to 1 pound of beef. It would only make sense to turn that wheat right into products you could eat directly, like bread. It would go 12 times further or feed 12 people rather than 1. These are going to become important ratios soon.

I'm not saying you have to become a vegan who eats no animal products at all, including dairy products and eggs; I'm just suggesting that you should begin moving your diet towards a more plant-based one. Start by having some meatless meals every week. This is easy for breakfasts and lunches, and then think of some simple dinners you already enjoy, like macaroni and spaghetti. Then try making some new meals like Chinese food where you make your own rice and maybe purchase some vegetable egg rolls or spring rolls. Start adding these gradually. Then try some of the outstanding meat-substitute products that are out there, like veggie nuggets and veggie cutlets. There are exceptional products on the market that use plant proteins and things like wheat gluten and are very hard to distinguish from the meat dish they replace. Try it with your kids. Next time you have hot dogs try some good quality veggie dogs but don't tell them. It's pretty hard to tell the difference anymore. Next time you feel like a burger, try a veggie burger. So much of the enjoyment of that burger comes from the toppings and bun that once you get it loaded up, it's often tough to tell what the main protein part of it is anyway.

These wonderful new meat substitutes are what I would call transition foods. They're allowing you to still have the things you're used to traditionally eating, but they're plant based. They're better for your health and better for the health of the planet, since all that grain is saved from being fed to animals. Gradually you're going to get used to these new tastes. You should also try more meals that you used to add meat to, like spaghetti sauce, and add soy-based "ground round" instead. Once you've done this for a while, switch to a Bolognese or vegetable red sauce or try pasta primavera with fresh vegetables chopped in. Then you'll be ready for those meals that break free of "the slab of meat as an anchor" type we all grew up on, and you can liberate your plate to celebrate the wondrous variety of plant-based meals.

This process will take some time. At first you might not want to disrupt special occasions which are focused on meat, such as Thanksgiving. But many people have come to regret that a creature must die in order for them to celebrate the harvest and have switched to some of the many new traditions such as making a veggie loaf with gravy or buying commercial products like Tofurky®. By adding all the standard mashed potatoes and cranberry and vegetables that traditionally adorn the table, switching to a tofurky is an easy progression.

It's important to realize as well the large-scale benefit of your movement towards a plant-based diet. First, far fewer resources like water and fertilizer have gone into your plant-based diet. Second, far less methane is produced. Cows in particular produce a lot of methane. Methane is 20 times worse as a greenhouse gas than carbon dioxide, so the less meat you eat the less methane your diet creates. I'm also suggesting that as you start producing your own food it's going to be way easier to grow enough potatoes to feed your family than beef.

I have to walk carefully in this minefield of dietary recommendations. Many of my amazing neighbors are beef farmers. As factory farms have come to dominate the agricultural landscape, my neighbors farm very sustainably. We live on a sandy soil left over from retreating glaciers, making it difficult to grow demanding crops like corn on a regular basis. So much of the cleared land is used for hay, which is fed to cattle. The cattle aren't penned up in stalls but get to wander in the fresh air. The problem is that farmers are being paid the same price for beef that they were 25 years ago. They raise cattle on a sustainable scale but it's very difficult to earn a living. A local group is now actively investigating biomass crops such as switchgrass. This is a perennial grass that will grow on marginal land and is a very efficient converter of solar energy into biomass. The switchgrass would be harvested and pelletized to be burned in a pellet stove. It appears that with North American natural gas supplies dwindling, my neighbors may be able to make more money growing fuel than food.

Whatever happens in the future, land that grows things is going to be very much in demand. Whether you're growing your own food or having someone else do it for you, you'll be able to get the best return on your time or money by focusing on a plant-based diet. Perhaps you'll save meat for special occasions like birthdays or for when family members you don't see often come to visit. For much of human history, meat, which was hard to come by, was reserved for just such times. It took more resources to produce and therefore was more "expensive" in labor than the traditional plant-based foods. Even though farmers who produce meat are not getting rich, meat is still relatively expensive to buy, especially on a weight comparison.

According to Planet Green, fruits cost an average of 71 cents per pound, vegetables cost an average of 64 cents per pound, and beef costs an average of $4.15 per pound (http://planetgreen.discovery.com/home-garden/save-money-by-eating-more-vegg.html).

You could serve people pasta with vegetables and garlic bread and a salad for the price of a 2-lb steak. By moving your diet to a more plant-based one you'll get more bang for your buck. As you start changing your diet, record how much money you're saving and be sure to put that money into your "Green Energy Account" to get your home more independently powered.

One of the nice things about a plant-based diet is how much easier it is to store foods. I know what you're thinking. You can buy a side of beef and store it in your freezer. That's true, but remember your freezer runs on electricity and requires continuous financial input to operate. And if the power goes off and you haven't installed your renewable system with battery backup, that meat can go bad very fast. Many plant-based foods don't require a freezer. Rice and pasta will store a very long time without the additional expense of electricity.

In the spring of 2008 as oil rocketed to $147 a barrel, the world experienced its first food shock in a long time. The price of all commodities went up and this included food products. Rice in particular became very expensive and in short supply. Asian countries that used to export rice began saving it for their own populations, which compounded the problem. As North Americans of Asian descent started realizing that there were shortages of rice elsewhere in the world they started purchasing extra, which caused supplies on the west coast to grow tight. Some stores even implemented restrictions. Big box stores in particular looked at the purchasing history on loyalty cards and if you didn't have a history of purchasing three 20-pound bags of rice a week, they weren't going to let you start now. This is one of the downsides of using stores that track your purchases.

It's a good idea to start building a supply of food in your home to help you the next time there are shortages. Many experts in the food industry are suggesting that it isn't **if** it happens again it's **when**. A root cellar is an excellent place to store vegetables, but it wouldn't be recommended for most of your food. You should start with a "pantry" or place to store dry and canned goods. It should be cool, dry, and dark. If you have a large closet that's not being used it's a good place to consider for your pantry. If all your closets are full it's time to pick the one that's the most convenient and best suited for food storage and have a garage sale with whatever is in there now. Storing food in the basement is only a good idea if it's not too humid down there. Basements tend to be cooler than upper floors but the humidity will definitely affect how well some foods last. While you may have an extra closet on the top floor of your house it may be too hot. Things like pasta won't mind the heat too much, but over time the foods stored in cans and any food with oil in it, like a rice kernel, will deteriorate faster. So the ground floor is best if possible. The closer to the kitchen the better.

The average temperature in your pantry should be above 32°F (0°C) and below 70°F (21°C) but remember that the cooler the storage area, the longer the retention of quality and nutrients. It should be dry, with less than 15% humidity, and food should not be stored on the floor. Try and have the lowest shelf two feet off the floor. You should also try to make sure insects don't get in, and have a mouse and/or rat trap if you don't think you can keep rodents out. If you have an abundance of large plastic storage containers, putting the items inside these is an excellent defense against pests, especially with things that are easily chewed into like pasta and rice that come in bags. It's also best not to have electrical equipment such as freezers, furnaces, and hot water heaters where you're storing your dry goods because they will produce heat, which you want to avoid.

Start with canned goods and things you know will keep. Pasta in bags will last a long time. Rice as well. Canned goods should last years. As much as possible you should avoid packaged goods like cookies and crackers because they'll get stale fast. You should start by picking up things on special. If you're at the grocery store and pasta and canned pasta sauce are on sale, buy what you would normally buy for your day-to-day consumption and then buy two or three more for storage. Look for specials on canned soups, canned fruit, canned vegetables, dried beans and lentils, and bagged rice and pasta. These can form the basis for your pantry. Then start to add those things you think will help enhance your menus, like sauces and spices, which last fairly well. Include non-essentials which you enjoy like salt, sugar, tea, and coffee, because they'll be nice to have if there are shortages.

Shopping like this will seem strange at first. Many of the things you'll be stocking up on are the sorts of things you might have purchased when you were filling up an extra bag during the holidays for the local food bank. You'll be buying things you might not regularly eat, such as canned fruit and vegetables. We've had the luxury of fresh produce 12 months of the year for so long that it's hard to remember ever having had to rely on canned versions of these. I'm not suggesting that you're necessarily going to shift your diet to canned versus fresh vegetables; I'm recommending that you start building up a three- to six-month supply of food that you have access to if you need it. It doesn't necessarily have to be a food-shortage crisis either. In the financial section (Chapter 19) we'll discuss how large a savings fund you'll need in case you suffer the loss of an income. Part of your defense plan can be to dip into your pantry during the transitional time until you can replace that lost income. Remember, the strategy is to start filling up the pantry with items on sale. So not only is that meal of dried fusili and canned Bolognese sauce less expensive than the fresh version and way cheaper than the restaurant version, since it was also purchased on sale it is even more affordable. You're protecting yourself not just from food shortages but from income disruptions, so stocking up the pantry is an excellent financial strategy.

Once you get your pantry stocked up you can start a maintenance program. As you add foods, label them with the date of purchase. Then, next time that item is on sale again, you can take the older item out to consume soon and put the newer item in. This is the same strategy that grocery stores use. They pull the older inventory to the front of the shelf and put the newer stuff in behind to make sure that the older items sell first. How long the items are going to last in your pantry is going to depend on how cool and dark and dry your pantry is. It's also going to depend on your comfort zone. We're all used to the "Expiry Dates" on some items we purchase. Some items like canned goods will have a "Packed Date" which will tell you when it was actually put in the can.

From a food-safety standpoint it looks as though storing these types of dry goods is pretty safe, because it's very difficult to find a govern-ment-sponsored website that gives you any guidelines. However, there are dozens on the safe storage of meat and dairy products. Apparently these are more problematic.

The following list deals with emergency storage for situations like a hurricane or severe snow or ice storm. I have not included readily perishable items that you would store in your fridge like meat and dairy products. This is for your long-term pantry storage.

Foods Recommended for Storage

- Water: one gallon per person per day for drinking, cooking, and personal hygiene
- Ready-to-eat canned foods: vegetables, fruit, beans, meat, fish, poultry, meat mixtures, pasta
- Soups: canned or "dried soups in a cup"
- Smoked or dried meats, commercial beef jerky
- Dried fruits and vegetables: raisins, fruit leather
- Juices: vegetable and fruit, bottled, canned, or powdered
- Milk: powdered, canned, evaporated
- Staples: sugar, salt, pepper, instant potatoes and rice, coffee, tea, cocoa mix
- Ready-to-eat cereals, instant hot cereals, crackers, hard taco shells
- High-energy foods: peanut butter, jelly, nuts, trail mix, granola bars
- Cookies, hard candy, chocolate bars, soft drinks, other snacks

The following chart from Colorado State University gives approximate times for food storage: (www.ext.colostate.edu/Pubs/emergency/fdsf. html)

Optimum Length of Storage for Quality and Nutrition

Fish, canned	18 months
Canned potatoes	30 months
Dehydrated potatoes	30 months
Canned fruits and vegetables	24 months
Canned fruit juice	24 months
Canned vegetable juice	12 months
Pickles	12 months
James and jellies	18 months
Rice, dried	24 months
Cornmeal	12 months
Pasta, dried	24 months

Cold breakfast cereal   12 months
Prepared flour mixes   8 months
Packaged dry beans, peas, and lentils   12 months
Canned evaporated milk   12 months
Dry milk products   24 months[1]

Canned foods keep almost indefinitely as long as cans are undamaged. The question is the quality of the food when you finally open it. In certain situations most of us would be quite happy to have some food that is safe, even if it's a year or two past its optimum storage time. I would suggest that the optimum times listed above are a best-case scenario for maximizing the nutritional value of the stored food. If you needed to eat and the food in the can had been there for four or five years, you'd probably be happy to eat it. This is a "beggars can't be choosers" scenario. The fact that most guides suggest that rice and dried pasta last two years tells me it will last much longer than that. It will simply be important for you to have a system of labeling so that as you put new items into the pantry you can take older ones out. I recently ate some canned mandarin oranges I put into my pantry four years ago and they tasted great. I believe that much of the labeling you see in terms of the storage life of products is based on product liability, where the producer does not want to risk someone getting sick from a product that has been stored too long. The longer that product sits unused, the greater the likelihood that it will experience less than optimum storage conditions like extreme and extended heat waves that will accelerate its deterioration.

From a cost-savings standpoint you should also look at purchasing some of the items for the pantry in bulk. This will save you money but will require a bit more time and possibly some expense in terms of how you store it. The advantage of purchasing things like flour and rice at a grocery store is that they come in a fairly well-made bag. If you buy in bulk you'll have to decide how you're going to store it. Mason jars are an excellent way to store things like dried fruit and beans, but you can avoid this expense by just washing glass jars that you would normally recycle. Start looking at what you're recycling and see if you're actually throwing away excellent storage containers. Plastic peanut butter jars are good because they have a screw-on lid that creates a fairly airtight seal and ensures that pests stay out. My favorite containers come from the brand of Caesar salad dressing that we purchase because they're glass but have a great plastic lid. With scary chemicals like BPA (bisphenol A), which is an endocrine disruptor in plastic, using glass wherever you can is a good idea, especially if what you're storing in it is acidic and may tend to leech chemicals out of the plastic.

So get creative. If you're buying in bulk to save money you'll need to think "outside of the box" (since it won't come in a box) to come up with the most cost-effective way to permanently store things. If you live in a neighborhood with curbside pickup of recyclables, once you find that favorite plastic peanut butter jar or whatever works for you it may be worth a quick walk around the block after dark or early in the morning to load up. Free storage containers! How great is that!

Fresh fruits and vegetables are extremely healthy and a huge luxury when you're eating them out of season. One of the keys to saving money and preparing for a time when there will be less diesel fuel for trucks to haul fresh produce from the south year round is to start to eat in season. Most of us have come to realize that a strawberry picked early in California or Florida and shipped to a place where there's snow in February is not as tasty as one picked at the height of the summer and shipped a short distance fresh from the local farmer. While the quality of those winter strawberries is getting better and better, is it really normal to eat strawberries in the winter in the north?

I would suggest one of the best ways to prepare for the inevitable changes to come would be to start eating as our ancestors did: in season. That means having strawberries every day for four weeks in June until you're sick of them. Then you move onto raspberries, then peaches, then apples in the fall and so on. By the late fall you'll be pretty much having just apples and pears, and as winter sets in your apples will be in pies and sauces.

It's probably time to start working on your canning skills. Many of us grew up with parents and grandparents who canned, but as we got busier with two-income families, and as lower energy costs made buying food so cheap, it didn't make much sense to can anymore. But those days are over. Canning and preserving are a lost art and one that you'll have to work on to get good at. Canning also adds a certain degree of danger to your food because if you don't do it properly your canned fruits or vegetables can cause botulism. While many of us are getting quite comfortable with the idea of injecting "botox" into our faces, eating food contaminated with it is another game altogether and one that can be fatal.

There are some excellent books about canning and preserving you may want to start with. You may also want to find a local course on the subject. It's always easier to learn from someone who's good at it. Maybe it's time to have grandma come and stay at your place for a couple of weeks at the peak of canning season and you'll have an in-house coach at your disposal. There's nothing like experience in this department.

If you are going to start canning consider using a pressure canning system. Our experience with canning has been that it is extremely energy intensive. The amount of energy you use for boiling, whether it's electricity or natural gas, is quite staggering. I marvel when I'm at the grocery store at the price of food, especially canned goods. Once you've done it yourself you get an appreciation for how cheap energy has contributed to our high standard of living. The company doing the canning on your behalf has to purchase energy for the process as well as purchasing the fruits and vegetables to begin with, as along with cans or glass, labeling, and cartons for shipping. The list seems endless and the final price to the end consumer mind-bogglingly cheap. This is just another way in which my research into peak oil and energy has made me grateful that I live at the time I do. Walking up and down a grocery aisle has become a time of wonder for me. Michelle eventually just ignores my ranting: "How can they possibly make a can of beans so cheap!" Now that I think of it, it's not just in the grocery aisles that she's stopped listening to me.

Using a commercial pressure canner ensures that you get the temperature to 240°F (116°C). This is the minimum temperature required to destroy botulism spores. Like a pressure cooker, a pressure canner will reduce how much energy you have to put into the canning process, which will save time and money.

Many of the books that have been written over the years for back-to-landers, those who want to get back to the simple life, always stress canning as a way to save money. Lots of money. I'm not really convinced in this day and age that that's the case. I won't argue that it might come to pass, but when I look at how long it's taken us to can things and how much propane we had to burn to do it, I'm simply not convinced that there are huge cost savings today. This is why I emphasize stocking your pantry with commercially canned goods now. You know they have been canned properly and someone else has had to spend all the money on the energy to do it. I think it's good to have the pressure canner and experiment with it to develop the skills to do it in the future when you have to, but as to it offering significant cost savings right now, I'm not convinced we're there.

A freezer is an excellent alternative or supplement to your canning.

This year we invested in a 10-cubic-foot freezer. While this may not sound like a big step, for someone living off the electricity grid it is. We had to evaluate how much additional load the 353 kilowatt-hours per year represented and determine if we could afford the electricity from the finite amount we produce with our solar panels and wind turbine. With our recent upgrade to more solar panels we felt we could. We also decided to put the freezer in our basement. Our house, which was built in 1888, has a concrete basement which is not heated. So it's more like a crawl space and stays very cool in the winter. Our feeling was that since the air temperature is close to zero the freezer wouldn't have to work that hard. And so far it appears to be the case. We haven't been able to notice any significant difference in our electricity loads.

If you were on the grid and paying 12¢/kWh, the electricity to run the freezer would be $42 (353 kWh x .12¢ = $42.36). We really loaded our freezer up with vegetables last summer. We froze beans, peas, basil, cauliflower, broccoli, and tons of tomatoes. Now on a cold winter day we take out a bag of tomatoes, put them on the woodstove in the morning, throw in some garlic and basil, and let it simmer all day, filling the house with wonderful smells. And their amazing flavor is like having a bowl full of summer. The tomatoes didn't go through a lengthy heating process by being canned; they were just put in bags and frozen, so I think we preserved some of the healthful benefits of freshly picked tomatoes.

Another process you might want to look at for preserving food is dehydration. Dehydrating food can be as simple as slicing it and putting it in the sun on a warm day. What you are attempting to do is remove the water through evaporation. The warmer and drier the air the better. To speed the process you may want to look at building a solar dryer. We have a solar oven, and during the hot sunny days of August we often don't need it since we have more than enough electricity for all our cooking. So we started putting tomatoes in it and it was amazing how quickly they would go from the garden to "sun dried," usually in one hot day. I also used one of my original solar oven boxes to supplement my output. The challenge with these is air movement, since you want to remove that moist air and keep hot dry air entering. Most models have a central drying shelf with a glass cover that creates a greenhouse effect, with a vent system at the top and bottom to bring hot dry air in at the base and vent the moist air out the top.

As energy and food get more expensive it's going to make more and more sense to take the time to build a solar food dryer and dehydrate fruits and vegetables from your garden. You will need to determine when the time is right for you. Building a solar food dryer may only require $50 in materials, but it may take a couple of days, or even a week if, like me, you're not a natural-born carpenter. If you still have a full-time job where you earn a wage, that investment in time may not make sense. If you are working part-time or are between jobs, it might make complete sense to build your own food dehydrator. Most people with gardens know that when the harvest comes in you'll inevitably find that you've planted more food than you can consume, so it just makes sense to have a low-energy way to save it.

You need to come up with what works for you in terms of how you save and store food. Find the best combination of purchasing commercially produced products, canning, freezing, or dehydrating produce from your garden, or buying from a farmer in season when food is at its cheapest. As food and energy prices change and your employment and income situation changes, you'll have to find the right combination that works for you at any given time. I would suggest you have the equipment and skills you need to exercise any of these options to varying degrees as things change and evolve in your life.

The "no-outside-energy-required-at-all" storage option for some fruits and vegetables is a root cellar. This may sound like a very old and rural concept, but a root cellar is an excellent, low-energy, very cost-effective way to store produce. Whether it's the bounty from your backyard vegetable garden or food purchased from local farmers in season, a root cellar is an excellent way to save money, become more independent, and reduce your carbon footprint. Potatoes and apples grown locally and stored with no refrigeration have a much smaller impact on climate change than those driven across the country or flown in from New Zealand or South Africa.

A root cellar can take many forms but the key is that it be cold and dark and humid. Ideally the temperature should be just above freezing to about 40°F (4°C). The humidity should be from about 80 to 90%. Knowing how dry the air in your home gets in the winter, you can see this is going to be a bit of a challenge. There are some excellent books listed in the appendix on building a root cellar. One option is to build it outside by digging it into a hill or by simply digging a hole, lining it with concrete blocks or stone, putting a roof on it, and covering it with soil. If you're living in suburbia this may not seem like a normal thing to do, but sometimes desperate times call for desperate measures. I can't think of a better way to get to know your neighbors than when they come over to ask you if you're putting in a pool or if that's a 1950s-style bomb shelter you're digging.

Older homes often have unheated basements which are an excellent starting place for a great root cellar. A dirt floor is even better, not only because the floor will get lots of dirt dropped on it anyway but because it will help regulate moisture. If it appears that the cellar is getting too dry in the winter, you can simply pour some water in the dirt so that it can evaporate out later. You can also do this on a concrete floor or use a bucket of dirt or gravel to get humidity into the air. If you have a finished basement then you need to build a small room, preferably on the north or east wall where it will be coldest and preferably with a small window. Or if you have an unfinished basement that you may finish, when you put in the root cellar insulate the walls to the inside to keep the heat out and leave the outside walls uninsulated to allow cold air to enter. There should be no ductwork or heating sources. The goal is to provide a way for warm air to leave the root cellar and cool air to enter. If your basement is naturally cool, leave one opening on the basement side and place one vent high in the outside wall for warm air to leave. This way the cooler air will have to travel across the root cellar and create some air circulation.

The root cellar needs to be dark to prevent potatoes and onions from sprouting prematurely. Try and seal up all the cracks to keep out rodents and pests, and make sure you have a couple of mouse traps baited at all times. Don't store any of your canned goods or preserves in the root cellar because it will be so moist it will cause any metal to rust. One compact fluorescent light bulb is all you'll need for light when you enter the room, and make sure you use a ceramic base and outside wiring because it will be cool and damp. If the room has a window you'll want to cover it with a heavy blind to keep the daylight out. One nice design I've seen takes the space of the window and has it half covered with a plywood vent that extends all the way to the floor and brings the cold air down to the floor of the root cellar. Above that is another vent for the warmer air which is higher in the cellar to exit, keeping a good circulation going.

Next you'll want to build shelves to hold all those fruits and vegetables. Some people like to put things like potatoes and apples in wooden crates. You can store potatoes in dry sand or peat moss in plastic buckets. I've found that wire mesh waste baskets are excellent for onions because the air circulates completely around them and I can hang the baskets up to get them off the ground. Squash and larger vegetables can just sit on shelves.

While root crops like potatoes like to be stored in a dry medium, carrots like a bit of moisture, so I store them in damp sand or peat moss. One of the keys to maximizing how long items will store is to ensure that you pick the best fruits and vegetables at harvest time and handle them as carefully as you can. Bruises and blemishes will hasten rotting and you'll often notice that this is where mould starts.

When you're harvesting potatoes or sorting apples, store the nicest-looking ones with the least signs of damage, and eat the other ones first. As the fall and winter progresses keep your eye on things and try and eat the items that look as if they may be starting to go. This is always is a good excuse for a nice soup or stew. If some of the skins on the potatoes aren't looking great just make a big pot of mashed potatoes rather than baking them.

Different fruits and vegetables are going to have different optimum temperature and humidity levels, and you cannot always control these perfectly. Every year will be a different experience with your root cellar and you just need to go with the flow. Ideally you should be harvesting potatoes when it's dry outside, but if August turns out to be wet you may not have a choice. That year your potatoes and turnips may not last as well, but the squash and apples will do great. Your diet will start to echo the health of what you've got stored. Late in the fall you should eat more of the items that won't last as well like leeks, turnips, and beets; well into the winter you'll still have lots of potatoes, carrots, onions, and some apples.

The majority of the vegetables I store in my root cellar are potatoes. I love potatoes and they are easy to grow and store and provide excellent nutrition. If you had to you could pretty much live on potatoes. Oh sure it would get a little boring, but the reality is that you can grow a good chunk of the food you need in a suburban backyard if you concentrate on potatoes.

Potatoes are rich in carbohydrates, which makes them an excellent source of energy. They have more protein than human breast milk, and since protein needs are greatest when a newborn is doubling weight every six months, we can obviously get all the protein we need from a potato. The amino-acid pattern of the protein in a potato is well matched to what humans need. Potatoes are very rich in many minerals and vitamins, providing one-fifth the potassium requirement, and are particularly high in vitamin C. A single medium-sized potato contains about half the recommended daily intake of vitamin C, so if you're having a crisis of conscience about that morning glass of orange juice that is trucked from the south to your breakfast table, don't worry; the potato has you covered. The United Nations declared 2008 the International Year of the Potato because, "The potato produces more nutritious food more quickly, on less land, and in harsher climates than any other major crop—up to 85 percent of the plant is edible human food, compared to around 50 percent in cereals."[2] Take this to heart and start making the potato a big part of your future food strategy.

Your root cellar will help to convince you just how amazing nature is. Our root cellar is actually the old water cistern under the kitchen. It has one-foot-thick concrete walls and stays just above freezing all winter. It is completely dark and yet by late March the potatoes are starting to sprout eyes, the onions are starting to send up green shoots, and the carrots are sending out feathery green growth on top. Somehow, even in the cool inky blackness, they know it's time to start the cycle all over again. And when those potatoes get to the stage where they're too wrinkly and the eyes are long enough, I'll take them out to plant in the garden and each potato will produce another eight to ten potatoes for next year. Free potatoes! After the initial purchase you'll have free potatoes for life. We continue to add new breeds just to keep a good variety to help deal with varying conditions, but growing much of your own food and storing it and harvesting the seeds for next year allows you to significantly reduce your food bill and makes you much less dependent on others.

I cannot emphasize enough the importance of having your own garden, but depending on space and other factors you may not be able to grow as much as you'd like. An excellent option is to join a CSA or Community Supported Agriculture. In a CSA you and a number of other families purchase a share in a farmer's harvest, in advance. I know, it sounds crazy, paying for something you don't have yet. But it takes away so many of the variables that make it tough for small farmers to make a living. It gives them money up front for seeds and they get paid for their time rather than having to wait until they sell the harvest. You as a member of the CSA also have a greater stake in the actual process of growing food. Your share of the harvest depends on what grows well. Some years you'll have tons of some items and very little of others. This is how it is when you grow food. Conditions change. Some vegetables do well, some don't. The megasuperstore has given us the mistaken belief that every shelf is always full of every possible fruit and vegetable. Cheap oil has allowed us to ship produce from wherever in the world it is growing well, but that's about to end.

So belonging to a CSA lets you experience a diet like the one your grandparents had. It also gives you a connection to the most important person on the planet. Not a lawyer or accountant but someone who actually grows food, something we're all pretty dependent on. CSAs have created a food production model that is also helping many younger farmers get into the business of growing healthy, local food. The traditional agricultural model requires massive capital to purchase land, equipment, and fossil-fuel inputs. The CSA model allows someone with a passion to grow food sustainably to earn a living. Some CSA farmers will also require or provide the option of working on the farm to offset some of the expense.

If you can find a CSA like this, join it. Set aside every other weekend or some weeknights or however much time you can afford to work with this farmer. First, there is nothing like the joy of hard labor nurturing food. In this case you'll get the added bonus of learning from an expert. It may seem as if you're just being assigned tasks and following orders, but you'll be learning what goes into growing food and there's no better way to put the joy back in a meal than to understand the toil and sweat that went into it. And make sure you make a nuisance of yourself picking the farmer's brain as to what she's doing and why. Why am I planting the beans here? Why am I watering the carrots and lettuce but not the garlic? The more knowledge you can soak up from this person the better your own garden will be. You'll also be adding to your knowledge base in case this is something you want to try yourself. There is one new reality in the economically and fossil-fuel challenged future and that is that more of our time and resources are going to be devoted to growing food. More human capital, more human labor, and more human time and ingenuity, because we simply won't have the gas and diesel to displace all the labor. Becoming an expert at food production will be a great skill set to have in the future.

The website Local Harvest (www.localharvest.org) is a great resource for finding a local CSA or farmers' market in your area. The site also lists local farms and Co-ops in your area. Food co-ops are an excellent way to save money on your grocery bill. You belong to a buying group, which increases the group's purchasing power and reduces its costs. Co-ops traditionally have been a means for people to purchase organic food, which in the past has been much more expensive than traditional food. Often you'll buy cases or larger quantities, which you may want to get other members of the group to split with you. Food co-ops are an excellent way to try and cut out the middleman of the grocery chains and get more money to farmers and smaller producers. They can also save you money on organic foods. According to the USDA, in the current economic model farmers end up with less that 20 cents out of every dollar spent on food; any way you can increase that is good for sustainable agriculture.

As discussed in the "Where to Live" chapter, one of the advantages of living in a densely populated city is the opportunity for food to be delivered centrally. So make sure that if you live in a city you're well versed in where and when all the farmers' markets are. Start to frequent them and make sure you take the time to get to know the farmers you're purchasing food from. Find out where they farm and what kind of farming they do. You may also want to find out if they'll sell to you at better prices if you order more or if you pick it up. It may be worth a drive out to the countryside in the fall to load up your car with potatoes and apples and carrots and squash and all those things you need for your new root cellar. If the farmer doesn't have to haul it into the city she may be happy to let it go at a better price.

If you continue to have animal products in your diet, keep in mind that smaller farmers are much more likely to treat their animals well. The agri-food industry has forced many farmers to grow food on such a scale as to require animals be kept in confined areas and often have antibiotics routinely added to feed to reduce the potential for sickness and economic loss. Smaller farmers may not have the capital for that or may choose to operate at a smaller, more humane level. You may pay a bit more for their meat and eggs and dairy products, but at least you know that the suffering of an animal isn't part of your next meal. I have been a vegetarian for 20 years but still eat some eggs and dairy products. The local farmer who supplies our eggs raises them organically and they live the life a chicken is meant to live, outdoors pecking around in the dirt looking for bugs. John also raises larger animals for meat, and as he says, "My animals have a great life, then they have one bad day." If I were going to eat meat, this is the principle I'd follow.

The other huge benefit of getting to know farmers is that in the case of disruptions to the food supply, be they man-made or natural, it's always nice to have a connection to someone who is in the business of producing that which is essential for life. It's unlikely that the chain store will really care if you can't get enough rice or other essentials for your family. That's not the capitalist model. The network of local farmers you've come to know will probably be much more compassionate in the same circumstances. They may suggest that they need your labor more than they need your paper dollars in trade, but so be it. It may be a nice distraction from everything else that's going on to get back to the simple and joyful task of growing food.

One final option that you may want to consider for your pantry is freeze-dried foods, which will last much longer than most of the other techniques I've discussed. Freeze-dried foods are sealed to prevent the absorption of moisture and can be stored at room temperature for years. With the very low water content the microorganisms and enzymes that would normally cause food to spoil and degrade can't do their work. When you add water and cook them the flavors and smells are excellent. Anyone who has camped and taken freeze-dried food knows how wonderful a nice pasta primavera meal is after a day of backpacking or canoeing. It's like having a restaurant-quality meal in the middle of nowhere with a minimal amount of preparation.

The downside to freeze-dried food is the cost. The freezing, primary drying, secondary drying, and packaging require lots of energy and lots of time, so you have a corresponding increase in cost. From a soft landing perspective it is probably too expensive for most people. From a hard landing perspective it is an excellent idea for a long-term food storage reserve. The food will last a very long time and because of the low water content it tends to be fairly light and portable. Companies like Nitro-Pak (www.nitro-pak.com), Mountain House (www.mountainhouse.com), and Harvest Foodworks (www.harvestfoodworks.com) are a few of the companies that sell freeze-dried food. Be forewarned that there can be a fairly strong survivalist theme in discussions of freeze-dried foods, but this should not prevent you from considering investing in a supply of them to round out your pantry. Just remember to keep them in a dark and dry place.

For anyone with a newborn or expecting a newborn I have not provided any guidance about what formula to buy and how to store it because I think you should be providing your newborn with the food that nature intended: breast milk. It has a myriad of benefits on top of the fact that it's cheaper and storage isn't an issue. As long as mom is eating a healthy diet, baby is looked after. From an environmental perspective it reduces packaging and the energy costs of producing and shipping formula. From a health perspective I think it is infinitely better for babies. As is so often the case, returning to the way our parents or their parents did things is a better choice than those provided by corporations today. We purchased 24 cloth diapers for our daughters when our oldest was born. Twenty months later when the second came along the eldest was just about toilet trained, so both the girls used the one set and they're still useful as rags in the garage. Think about how many trees weren't cut down and pulped and shipped and processed and used once to end up in a landfill. Disposable diapers have a hugely negative effect on the planet and on your pocketbook. Ask for cloth diapers at the baby shower; then calculate how much you would be spending on disposables each week and take that money and set it aside. I'll bet you can purchase a significant number of photovoltaic panels with that money. Independence often involves choosing what's right for the planet and your pocketbook, not what's most convenient.

I've provided a roadmap here for an evolution of your diet. I'm suggesting you move away from an animal-based diet to one that is based on plant materials. Make the anchor of your meal a starch or carbohydrate like rice or pasta or potatoes, and then add your vegetables as a side. This diet has worked for centuries for a huge majority of the planet's population and will work wonders for your health. If you're having any problems getting over the "low-carb" propaganda of several years ago, just remember that this book is going to have you doing so many things physically that you used to use fossil-fuel energy for that you'll be able to eat all the carbs you want and burn them off.

Get a pantry stocked with six months' or a year's worth of food and when you can, do it with items that are on sale. This makes for a fallback in the event of job loss or food disruption. Then build yourself a root cellar and stock it fully every fall. If you can grow the fruits and vegetables you store, excellent. If you can't, join a CSA or get to know some local farmers who'll sell you large supplies in the fall to help you stock the root cellar.

A plant-based diet is going to be much easier to maintain in the future, and having a pantry and root cellar fully stocked with a reserve is going to greatly increase your family's independence and allow you to weather storms as they arise. It makes complete sense economically because this strategy is going to save you a significant amount of money over your old packaged and processed food diet. It's going to be much healthier for you and infinitely better for the planet since it decreases how far your food travels. And by eating the grains that used to be fed to livestock, you'll be significantly reducing the greenhouse gases that are permanently altering our atmosphere. You win, the environment wins, and your pocket book wins. FFFF–Food, Fuel and Financial Freedom start in your pantry! Bon appétit!

14 Gardening

A man without land is nothing
Mordecai Richler, The Apprenticeship of Duddy Kravitz

The typical North American diet is one rich in fossil fuels. From the natural-gas-based fertilizers to the diesel used in tractors for planting and harvesting and in trucks for transport, to refrigeration and packaging, the fossil-fuel inputs into our diet are enormous. With many geologists warning that the world has hit peak oil and North American natural gas producers telling us we're rapidly depleting our remaining reserves, the percentage of our incomes that we devote to food is increasing. Biofuels are competing for grains such as corn and soybeans, and climate change is causing droughts which are hampering grain harvests, driving stocks to their lowest levels in years. As all of these variables drive up the cost of producing our food, you should plan on spending more of your paycheck at the grocery store.

There is one way to offset some of this food sticker shock, and that's to start growing some of your own. Whether it's in your backyard or at a local garden rental plot in your city, it's time you took control of what's on your dinner plate. Gardening is phenomenally popular, but most North Americans focus on flowers, and while some flowers are edible, they don't add to your family's food security or help your monthly budget.

Tremendous media attention is given to the concept of eating more locally, like the 100-mile diet, where you try and offset some of the miles your food travels. The estimates vary, but you'll find that your average North American meal has traveled between 1,000 and 2,000 miles. Some estimates have it as high as 2,500 miles, which has a tremendous impact on the planet in terms of the greenhouse gas emissions created to get it to your plate. For much of the year fresh food being eaten in the northern States and Canada has had a long truck ride before it arrives at your grocery store.

So whether it's because your food budget is getting squeezed as the costs rise or because you think food costs will increase aggressively as we run out of the easy-to-find fossil fuels, growing your own food is a great idea. Michelle Obama planted a vegetable garden on the lawn of the White House, and the City of Vancouver, host of the 2010 Winter Olympics, plans to have 2,010 community gardens by 2010. Local and urban vegetable gardening is breaking out everywhere.

From a hard landing point of view the decision to grow some of your own food is simply not an option. You have to. This will be one of your sources of sustenance. And like so many of the other new activities you'll be engaged in, you'll find it incredibly rewarding as well. There is nothing like eating food that has come from the seeds that you planted in the ground. It's something that's uniquely human and something that goes back to the time when our species stopped being hunters and gatherers and started to use agriculture as a means of self-support. You're going to really enjoy "The 100-Foot Diet" as you bring your homegrown vegetables into your kitchen.

I have been gardening for more than 30 years. When I was 16, I decided to put in a vegetable garden. I cannot tell you why. As a 16-year-old in 1975, I was interested in girls, and cars, and music, and school, but something told me to grow vegetables. I had no experience with it. My parents had planted flowers at some of our houses, but I'd never spent any time with tomato or potato plants. I just wanted to grow vegetables.

We were living in Burlington, a suburb of Toronto, Ontario. Like so many suburban developments, all the topsoil had been scraped off the land before the houses were built, and we were basically left with solid clay subsoil. It would have been great for making ceramic pots, but for growing things it was less than ideal.

But I used my shovel, turned over the grass, put in some peat moss and planted my vegetables. I can still remember the condition of the soil after a few weeks. It looked like every image of a drought-stricken patch of land you've ever seen on the cover of environmental books. It was depressing. It didn't matter how much I tried to keep the soil loosened up, every time it rained the clay congealed back into a clumpy, dense mess.

Luckily we moved the next year to a home in Belleville, in Eastern Ontario, and even though it was in a city the soil was amazing. It was a hundred-year-old house, built in a day when they left the soil alone. I put in a garden and it thrived. In fact, I was so enthusiastic about gardening that I got a job at a local garden center. After moving out of the house I lived in various apartments, where I always found a local garden plot to rent. In our apartment back in Burlington years later, Michelle and I rented a plot at Bronte Creek Provincial Park. It was about a 20-minute drive and a huge hassle, but I was still drawn to having a garden. I remember watching an older gentleman covering his potato plants with a white powder that looked like the DDT they used to douse POWs with in World War II. With our growing concern about the effects of these chemicals on us, we were thrilled to finally buy our own house.

Here I could finally have my own garden, away from any outside disturbances. The first garden went in the backyard, which boasted two huge black walnut trees. These amazing trees provided wonderful shade for our daughters to play in and kept our house cool on hot days in the summer, but we later discovered they poison the soil for most plants. So the vegetables were dismal.

I had gradually been turning the front yard into flower beds, so each year more and more of the beds ended up with vegetables in them. At first it was a few tomato plants, quite discreet and hardly noticeable. Then a few bean plants got thrown in. We had planted a peach tree in the front yard, and each year it got bigger, one year providing us with over 100 peaches! This was a small front yard, and eventually it was jam-packed with vegetables. Some people thought it was great and some people thought I had lost my mind. We had a home-based electronic business and my customers were immensely amused by my front yard. As long as I got their work done they tolerated my quirkiness. When my chiropractor asked if that was corn growing in my front yard I decided it was time to get out of the city and get some land. Some land for a big garden.

For more than 30 years I've been reading books about gardening and talking to lots of experienced gardeners, and from all of this I've learned that growing food is really easy. You'll have failures and you'll have successes, but if you keep at it long enough you'll figure it out. You need compost to build your soil, water to hydrate your plants, and time to weed and nurture your vegetables. That's all. It's not rocket science. All you really need is motivation and you'll have a great garden. I hope Part II on the challenges we face has motivated you enough to plant a garden. Now I'll share a few of the things I've learned over the years.

Getting Started

So where do you start? First you need some land. How much land? Well that depends. To begin, why not get started small with the intention of expanding. If you own a house, find a spot in the yard that you can spare. If you can't find any room to spare, something's gotta go. Maybe it's the hot tub, maybe it's the gazebo, but your priorities have just changed so it's time to get serious about food. If your pool takes up most of your backyard, it's time to dig up all the trees and shrubs so you can plant vegetables. The time to worry about your yard looking fashionable is over. If you still own a house you should be happy to have the opportunity to grow your own food.

If you live in an apartment start with some garden boxes. Hopefully your balcony has enough sun to get some tomatoes going. Another great option is to find a local garden plot that you can rent for a season. Many cities are turning vacant property into community plots so local people can grow some of their own food. These also make an excellent place to meet other people who may share your world view and to talk to and learn from more experienced gardeners. As I discuss in Chapter 9, living in a community is going to be a good thing during challenging times. A community garden allows you to get to know more people in your city and form a group. This always helps the isolation so many people feel in the city and will give you greater resources when things go wrong.

To find a local garden plot start with your local city hall and see if they can recommend a place. Put up a sign at the local recreation center, ask people you already know in the community, find a local environmental group as they'll often have a listing of these, use the Internet, do whatever you have to do to find one. The one thing you may want to check is whether soil tests were done prior to turning the location into a garden. The unfortunate thing about cities is that there may have been a commercial or industrial enterprise there previously, and the soil may contain lead or other chemicals it would be good to stay away from.

If you don't have room and can't find a local garden plot another option is to find a CSA (Community Supported Agriculture) farmer close enough to reasonably get there to work on the farm. This way you'll be learning and getting the experience as well sharing in the harvest. Can you take your bike on the local transit system and ride to the farm from there? Do you own a car and could you devote one whole day on the weekend to working there (and the next day to recovering?) Get creative, but get your hands in some soil somewhere soon!

Soil Preparation

Many people begin with land that is covered in grass. There are lots of ways to turn that grass patch into a garden. The most direct way is by using a shovel and turning over clumps of grass to expose the dirt underneath. If you're thinking about starting a garden next spring, and it's fall now, just turning over the grass should work fine. If you're a few months away from planting and there's no snow on the ground, lay down some old newspapers or spread out the cardboard that your new energy-efficient fridge came in and this will kill the grass over time. You might be concerned about contaminants in the newsprint ink, but many printers have switched to vegetable-based inks, and since most of us don't eat organically all of the time, I'm not too concerned. After a few months under papers or cardboard or even an old large pool cover you pulled out of the neighbor's garbage, the grass will be much easier to remove. If it's really dead you can use a hoe to pull it apart, but a shovel may still be required.

It's important to make sure you keep the soil that the grass is rooted in. That will be your best topsoil, and if you pull out the top four or five inches of grass and soil and discard them you'll be losing it. The topsoil will have the most potential nutrition for your vegetables and will be full of micronutrients and microscopic organisms that make soil healthy. Those organisms are going to break down compost and make the essential elements that plants need to access easily. If you have a heavy soil or healthy grass with a large root system which is taking a lot of topsoil with it, do your best to knock the soil off. Then pile the clumps of grass nearby so you can compost them and ultimately get all the valuable materials back over time. In much of my garden I turn over the grass in clumps and then use a multi-tined hoe and rake the clumps across an area that I've already prepared. Each time I grab the sod more soil falls out and eventually I toss the remaining grass into piles to compost.

If the area you're to garden was cleared before or has never had grass on it, you may be able to use a rototiller. This is a gas-burning machine and you may be loathe to use such a thing, but when you are just getting started it can save an amazing amount of work. I use a rototiller in my workshops as an example of the energy in a gallon of gas. Three tablespoons of crude oil represent the equivalent of 8 hours of human labor, so a tank of gas in a rototiller can replace a whole day of shovel and hoe work. I'm not recommending it; I'm just throwing it out there as a possibility. I wouldn't suggest you purchase a rototiller, but if you can find a used one or rent or borrow one in the spring when you need it, it can improve how much you can accomplish in a short period of time. If you have the time, then grab that shovel. I've already suggested that you scrap the health club membership, so this is one of the activities that will help you keep in shape. And long-term, as the price of gas increases and it becomes more and more scarce, it may not be possible to run a rototiller. If you've got some jerry cans of gas stored from last year, a rototiller is a good small engine to burn it up in so you can replace it with fresher stuff.

If you decide to use the shovel, purchase a file and sharpen the shovel blade. Most new shovels won't be sharpened, and if you're using a dull shovel blade it's going to make your job much harder. Also make sure you have a good-quality pair of workboots. Pounding down on the shovel with running shoes to push through densely matted grass is really hard on your feet. You need a boot with a very rigid sole, and if you're as dangerous with a shovel as I am a steel-toed workboot is recommended.

Rototilling over a patch of light grass or other plant material will break it up, but the grass will still be in the garden, so when you've finished use your cultivator or rake to remove the grass. This will keep it from taking root again and reducing your vegetable yield. If it's a large area of stubborn grass, rototill up and down the garden, rake and remove some grass, then rototill the garden side to side and rake again. It's really important that you try and start your garden with mostly soil and not a lot of roots and weeds that are going to have to be removed later. If you have access to a large amount of mulch like rotten hay or leaves from last fall you won't need the soil quite so pristine, since the mulch will keep the weeds down around your plants once you apply it. Weeds are going to be one of your biggest enemies as the summer wears on, so try and eliminate as many as you can right up front.

With your soil cleared the next thing you do is add compost. Compost is the key to a successful garden. If there is one thing you should get out of this gardening chapter it's the importance of compost. Compost, compost, COMPOST! I'm really serious. In all the books I've read and research I've done and market gardeners I've spoken to, compost is King. Compost is going to give your soil life. It's going to make it rich and provide plants with the nutrition they need for vigorous growth. It's also going to provide humus and the organic material it needs to help the soil retain moisture. The remains of grass clippings and leaves and hay and vegetable peels are all going to be in various states of decomposition and they will absorb water when it rains and release it back to the roots during dry spells. A good compost is the key to a healthy garden and good yield. There have been many times when I've spoken to lifelong gardeners about specific problems I may be having in my garden and invariably the solution is "more compost." They may just be saying this because they don't have any other pat answers, but it's always worked for me.

It doesn't matter exactly what is in your compost but you'll want a lot of it. It's great if you're composting all your kitchen scraps but these generally aren't enough. You need to supplement, and since you're growing a vegetable garden you need to supplement in a big way. Start by looking at all the natural organic material around you that ends up in a landfill. Some cities now have separate wet and dry pickup, but many haven't taken this step. The best you might get is leaf pickup in the fall, but this is a great place to start. Twenty years ago I lived in a city that had a crisis as its landfill was filling up yet didn't pick up people's leaves separately. In the downtown area where I lived, every fall there were thousands of green garbage bags full of leaves out at the curb to be trucked to prematurely fill up the dump. You could easily recognize them because a bag of garbage would have lots of pointy boxes and irregular shapes sticking out of it. A bag of leaves would be perfectly round and cylindrical with the indentations of leaves in the plastic. So on the night before garbage day I would set out with my wheelbarrow and load up 3 or 4 bags at a time and walk them back to my place. By the end of the season I would usually have about 100 bags of leaves in my pile behind the garage. My daughters loved it. By the next fall, after a summer of turning the pile over every two weeks with a hay fork, I would have a huge pile of wonderful dark, rich compost. The bottom layers would be full of worms that had come up through the soil to work their magic in the leaves, eating their way through and leaving dark rich castings to become part of the compost mix.

Eventually the city started picking up the leaves and asked residents to put leaves in clear plastic bags. This made my job even easier. Now many cities have asked homeowners to put organics like leaves in large paper bags that will compost down. Your neighbors are getting rid of a gold mine of organic material that you need for your garden. Ask them for it. I'll bet lots will even haul it over to your place as soon as they've finished raking just to get rid of it. Grass clippings are something you should be getting as well. With grass clippings, though, you might have an issue with pesticides. I always watched to see which neighbors sprayed their lawns and ended up with the little white warning signs. When I was grabbing grass on my garbage-night wheelbarrow runs I'd try and avoid those places. I even convinced a few neighbors to stop spraying so I could use their grass clippings, and they didn't seem to mind. Grass clippings are excellent for your compost. They are very high in nitrogen, which is one of the key building blocks of healthy soil.

Then it's just a matter of finding as many sources of diverse organic materials as you can. Is that local coffee shop throwing out coffee grounds? Why not provide them with some buckets that you'll pick up regularly. Perhaps that fruit market or even the grocery store where you shop has to dispose of fruit and vegetables that are no longer saleable. What about that local greenhouse/craft store that had the excellent fall display with all the corn stalks? Ask for them, since they may be going to throw them out. Or that church with the huge nativity scene with all those straw bales. "Can I have them when you you're through?" Crushed eggshells are a good source of calcium and your cabbages will like them, so ask the local diner with the busy breakfast trade if they'll toss their eggshells in a bucket for you to pick up after every busy weekend.

Get creative, but make enlarging your compost pile a priority. Once you start assembling these various materials you should start mixing them together. If you have many bags of leaves and grass clippings, hold some back. Take as large an area as you can find to make your compost pile. The days of your nice, neat little black plastic compost barrel are over. If you want to grow a lot of vegetables go big on your compost pile or go home. There are lots of designs for larger compost bins, but an easy one is wooden pallets. Stake them so that you have two sides and a back. As you build your pile some will fall through the spaces between the boards of the pallets but because they're open it allows air to get in. If you find two more pallets, make a matching compost area beside it and you can have two piles at different rates of decomposition, one that's garden ready and one that may not be ready for a few months.

Your compost should consist of layers from as many sources as possible. A layer from your kitchen composter, then six inches of leaves, then six inches of grass clippings, then an inch of coffee grounds, half a bucket of egg shells, then start over again. If you're in a rural area hopefully you can find some rotten hay and straw to include. Start mixing this in.

Every week or two, do some stretching exercises, then get your hay fork (with the long pointy tines, like the ones the villagers grab when they're going to chase monsters in the movies) and move the compost pile over three feet. You'll have to grab layers and sections at a time and heave them into the new pile. This will mix the pile up and get it decomposing faster. It would be good to have a garden hose handy and wet both piles as you work. The moisture will help speed things along and make a better environment for the organisms you want to attract to break down the materials. For the first month or so the pile won't look that different, but as the summer progresses you'll start noticing that more and more of the pile looks like soil. After eight or ten weeks of this you'll be amazed at how much it has decomposed and looks like those wonderful pictures you see on the side of the commercial composters with people putting in kitchen scraps at the top and beautiful black topsoil coming out the bottom. That has never been my experience with my big plastic composter. I find the materials sit and smell and turn into a big disgusting gooey mess. One of the main reasons for this is that the material doesn't get enough air. Anaerobic decomposition refers to the breakdown without oxygen, while aerobic occurs with oxygen. So if your compost isn't getting any air it will not break down very quickly and it will smell. If you're turning it over every couple of weeks it will be getting lots of air, and the fact that it's now a pile means air can get into it and keep any smells to a minimum. You'll know your compost pile is working well if it's getting hot. That heat shows you the microorganisms are working hard because heat is one of the byproducts of the aerobic decomposition process. So the more heat the better. If you're turning the pile over on a cool day in the fall and you see steam coming out of it, you've hit the compost pile big time! "You've gotten to the rotten zone!" Celebrate "Decay Day"! If you blog, write a post and boast how you toast your compost. OK, I'll stop now.

I realize that a large compost pile is outside a lot of people's comfort zones. If you get too much grass in there and don't turn it often enough it'll get stinky. I'm just suggesting that if you're serious about having a good garden this is how you're going to do it. You can always buy commercial fertilizers which will produce excellent results, but you're basically getting your soil addicted to fossil fuels. You'll need to keep adding this fertilizer because your soil will get lazy and so you'll have to keep using your after-tax dollars to keep buying something that adds much less long-term value to your soil than the compostable materials that your neighbors are throwing out, for free! If you do decide to buy things like bagged manure, keep your eye on all the garden centers, especially the temporary ones set up in grocery store parking lots. They'll often get pallets of these bagged products and by August will be ready to close up shop. The bags that are left may be beaten up, but this is when store owners offer them at a heavily discounted price, and those bags will give you a really excellent return on your investment of time. Take your duct tape and offer the retailer 10¢ on the dollar to take the broken bags. If you can borrow a neighbor's truck, maybe you can grab the pallets to build your composter with at the same time that you're grabbing the bags of manure. The store owners may be happy to get rid of it all in one fell swoop. They're going to have to move it when the season winds down anyway.

In my garden I'm looking to provide nitrogen and carbon. Grass clippings or any fresh green material, kitchen scraps, and manure are all high in nitrogen. Corn stalks, leaves, straw, and hay are good sources of carbon which also have some nitrogen as well as phosphorous and potassium that your plants need. When you buy a commercial fertilizer it will list these three items in order: N = Nitrogen, P = Phosphorous and K = Potassium (potash), or NPK. Typically a lawn fertilizer will be high in nitrogen to make your grass green so the ratio will be 21-7-7. A commercial garden fertilizer may be 6-10-4. The more you read about what combination of these is optimal for a garden the more intimidated you can become.

Obviously some plants will like more of one element and less of another. Some will require trace elements difficult to obtain from any given source. My approach to this has always been the shotgun approach. I spread as wide a path as I can in terms of what I put into my compost and garden and hope that I get what I need. This is not completely scientific I know, but there are so many variables with gardening and you can use up so much time planting and weeding and watering that sometimes I believe you can over-think this. If you can afford to purchase a commercial fertilizer and you've just moved into a new subdivision where all the topsoil has been stripped away, go for it. If you don't have access to a nearby horse farm for manure, buy some commercial bags of it. If you get lots of grass clippings in your neighborhood but not too many leaves because the trees aren't mature enough, make sure you don't overdo it. And when you have to be in one of the older sections of town in the fall, make sure you have an old blanket or tarp in the back seat so you can load up on bags of leaves at the curb. Try to get as big a variety of compostable materials as you can and you should be fine.

When we lived in an apartment we had a vermi-composter, a large wooden box lined with plastic that had "red wiggler" worms in it. We put our kitchen scraps in it and the worms did a fabulous job of breaking them down. Periodically we'd take the compost to my father-in-law's garden. When we moved into our house we put the red wigglers into the area where our compost pile would be and they thrived. Years later as I turned over the compost there would be masses of worms in it, breaking it down. There were handfuls of them! Even though our area experiences several months of below freezing temperatures, with so many bags of leaves piled in there in the winter the worms were all nicely insulated and back to work in the spring.

By the fall you should be able to start putting some of that compost on the garden you've been working on and turning over all summer. In the following spring take out some more and add it to the garden. Make sure you keep some of those worms you'll see in the compost piles and they don't all end up in your garden. It's great to have them in the garden, but on the scale of your garden they're best kept in the compost pile.

Manure is also an excellent soil conditioner. If you're starting a garden and you're in an urban environment and can afford commercial manure, go for it. Sheep and cow manure will come in plastic bags and will already have been composted down nicely for you. This is an excellent way to start building up your topsoil, which is the critical part of your garden. If you're in the country try and find a local farmer who has extra manure or a nearby horse owner who may have some to spare. You have to be careful with manure, especially cattle manure, because it may contain E, coli and other nasty organisms that can make you sick. If you can get manure find out how long it's been sitting. If you're unsure, pile it away from your garden for six months until you're sure it's safe. As with your compost pile, make sure it gets hot, as the heat kills pathogens that you don't want near your food.

I realize this section on compost is long and drawn out! Yes, it is. But there is a point to it. If you don't start with well-conditioned soil, everything else is a huge waste of time. You may enjoy some success for a while but the lack of soil conditioning will catch up with you and you'll never know whether those black spots on the leaves are a blight or just a plant in distress because its roots couldn't find the nutrition it needed in the soil.

Planting

One of the greatest things about gardening is the anticipation of spring, and one of the best ways to hurry spring along is to start some of your own plants indoors late in the winter. By March your local greenhouse will be in full swing planting flats of flowers and vegetables for customers to buy in April and May (depending on where you live). To save yourself some money, you should be doing it yourself. You'll get a better selection of seed types than the vegetable plants from your local retailer, which is one of the other advantages of starting early. Many people swear by their "heritage" tomatoes and claim that they taste infinitely better than standard commercial brands. Some people have started saving and swapping seeds with others to increase their variety. And there is nothing nicer on a cold winter day than sitting down with a seed catalog and going through the color photos of all those fruits and vegetables that you'll be able to grow next summer.

Waterproof seed trays are an excellent way to start seeds, especially if like us you'll be moving them on a regular basis. Some people go to the expense of purchasing grow lights because it's often hard to get enough sunlight in your home in the winter months for seedlings to thrive. These are a great idea. Living off the electricity grid we have yet to invest in them. We start our seeds in seed trays with a plastic cover, like a small greenhouse, and we move them from window to window during the day as the sun moves. The greenhouse lid keeps them warmer at night. Eventually as the days get warmer we can put the whole tray outside so that the plants can get full sunlight. This is called "hardening off" the plants so that it's not such a shock when you do finally put them into the garden. It may sound like a lot of work moving them to follow the sun during the day, but it also keeps you watching to see what's coming up and how the seedlings are doing.

Starting seeds indoors will be something you'll have to get a feel for based on what you like to eat, when the last frost hits your area, and how big your garden is. We've developed a hybrid system where we start some things ourselves and buy others from a local greenhouse. Some days we don't have the electricity to have grow lights going 24/7 so we know that things like tomatoes and pepper plants will not thrive and will be gangly plants that are too tall and weak. By the time we are ready to plant tomatoes in our garden, the commercial greenhouse has tomato plants that are thick and lush and compact. They might cost us $1.49 for four plants, but they are beautiful plants that thrive, so this is a tradeoff we're prepared to make. We do start some of everything and we do put some of our own tomatoes in the garden but we supplement them with greenhouse tomatoes to make sure we have a good mix. We generally start our plants in March and then start another set in April and more in

May. We do this so that we have plants at various stages to put into the garden. The challenge with purchasing commercial plants is that if you plant four broccoli and four cauliflower when you put the garden in after the last frost, when they mature you'll suddenly have more broccoli and cauliflower than you can enjoy. Some of it will end up in the freezer but we make sure than we have a number of plantings and have everything at various stages of maturity to extend the season.

Once the ground is warm enough to be worked you'll want to get out there and get the garden ready. Remove any plants that you left in from last year and weeds that didn't get pulled in the fall. Work some compost in and rake the soil to get it ready. Some things can be planted early, like spinach and radishes. We'll often have seeds self-germinate from last year and it's amazing how early they'll germinate and begin to grow. Sometimes there's still snow in certain parts of the garden and spinach plants will have started growing. Lettuce seeds are easy to save so I usually put in a row of these early. Peas can also tolerate cold weather and in fact I find they like a frost. It's always nice to get these early vegetables going to entice you into the garden after a long winter. There's nothing like that first salad where you use some of that early lettuce and spinach to remind you that spring is here.

There is much debate about how much room to leave between plants and rows. This is almost a bit of a city/country debate. There are many books about bio-intensive or "square foot" gardening that suggest you can grow huge amounts of vegetables in a very confined space. The books that discuss this are fairly technical in their description of how to do this successfully. If you have limited space I would urge you to learn more about this, but with the space I have it's not a technique that I require and so I am not going to go into it in great detail. I would suggest you start by following the seed package recommendations on how far apart to plant things. If you have limited space and the seed package recommends rows 3 feet apart, try making them 2 feet apart or even just 18". By harvest time the plants will have crowded out the walkways between rows and it will be harder to get around the garden. But by then your priority will be harvesting and eating those vegetables, so if the garden looks a little unsightly it will be a small price to pay. If your vegetables are this dense it also makes it harder for weeds to grow, so there is some benefit to having tightly grown plants.

My attitude about bio-intensive gardening is simply that plants need nutrition and water and there is only going to be so much of that in any given area of the garden. If you crowd too many plants into an area that can't support them you'll simply reduce the harvest. So you really have to be paying attention to do this well. If you're starting a garden without a good handle on the quality of your soil, I would recommend that you go with rows a reasonable distance apart. As you get more comfortable with your garden and you've been able to provide more compost and supplements you can begin increasing the density of your plants. After a few seasons you'll also have enough experience to be able to decide how much of everything to grow. After a summer of realizing that no one in the family likes kohlrabi and that a 12-foot row of beans grows more beans than you'd eat in two years, you'll be better prepared to scope out a strategy for intensive growing.

You should always rotate where things grow in your garden. This ensures that insects and diseases that remain in the soil are less likely to have an easy time getting to your crops. Each year I change the orientation of my garden with rows going vertically one season and horizontally the next. I make sure that my heavy feeders like corn are moved around each year. This also allows me to ensure that the walking paths are different each year so I don't end up with severely compacted soil.

One thing you should use to extend your season and save money is a cold frame. This is a small greenhouse you use to help start seeds in the spring and keep some sensitive crops going in the fall. There are lots of plans available but I built mine out of four windows I found at the local dump. They were old and have a wooden frame so I just screwed three of them together and put the top one on a hinge to open and close. I left the back open and have it against the concrete foundation of my house. This way the sun warms the concrete during the day and the heat is radiated back at night. A cold frame will keep plants from getting nipped by frost. It's also a great place to put your flats of seedlings after the sun goes down. When it's warm enough you can move them outside during the day to harden off and back into the cold frame at night to protect them. We also start some lettuce and spinach in our cold frame in the fall and it gives us salads until December. Growing in a colder climate as we do we try everything we can to extend our season. So keep your eye on your neighbor's garbage and always grab any old windows you see. If you have a few of them you can make a cold frame and then if you accumulate enough you may be able to make your own greenhouse someday. Since your garden is reducing the distance your food travels you are actually increasing your greenhouse glasses to reduce your greenhouse gases (loud groan)! I would like to thank my brother-in-law for that pun. If you disapprove contact him directly.

Your local weather or government agricultural office can provide you with a final frost date if you're not familiar with it. As you get closer to that date you may want to start putting in some cool-season vegetables like beets, broccoli, cauliflower, cabbage, and beans. If I'm going to make my rows ten feet long, I'll seed half of it a week or two before the last frost date and the other half a week or two later. Hopefully some of the seeds will germinate and will get a bit of a head start on the second planting. Some seeds need the soil to be fairly warm to germinate so I'll wait until our final frost date to plant carrots, celery, potatoes, and onions.

That frost-free weekend will tend to be my big gardening weekend when I'll plant as much as I can. I'll plant corn as well as my "vine" plants like squash, cucumbers, muskmelons or cantaloupes, and pumpkins. You have to be sure that the danger of frost has passed and the soil temperature is starting to get high enough for the seeds to germinate. This is also the weekend I'll pick up the bulk of my vegetable plants from the local greenhouse. These will be my real heat-loving plants like peppers, tomatoes, eggplants, and watermelon. I'm fortunate that I have an old concrete barn foundation on my property. This is a real advantage for these heat-loving plants in a colder climate like mine. Even though the danger of frost has passed, night temperatures can still be quite cool for a few more weeks and this slows down the growth of these plants. I built raised beds in the barn foundation because it has a concrete floor. I keep the soil in place with old cedar posts that I retrieved when the road crew replaced some of the guardrail posts along our road. They are just cedar trees and are not chemically treated. The beauty of the barn foundation is the thermal mass of the concrete. When the sun is out during the day the concrete absorbs heat, and once the sun goes down that heat is radiated back. The concrete walls also act as windbreak for some of those cool breezes and help nurture these fussy plants that act as if they need a Caribbean holiday to get growing.

Be careful with real heat-loving plants like peppers and eggplant. While I buy them with everything else on May 24th, I don't actually put them in until the middle of June. I find that if they're in the garden and the nights are too cool their flowers won't set fruit properly later on. I keep them outside during the day but keep them inside at night until I'm ready to plant.

Raised beds are a fine idea if you have a limited amount of space.

They allow you to grow fairly intensively without compacting the soil. The roots of your plants need oxygen and if the area around them is heavily compacted they may have trouble accessing it. With a raised bed you do all your walking around the outside of the bed, which leaves the soil fluffy and aerated. Since my garden ends up with well-worn walking paths, I like using my rototiller to loosen that soil up in the fall and again in the spring before I plant to ensure that the roots have an easy time making their way through the soil in search of water, nutrients, and oxygen. With a raised bed be really careful with the material you use for the sides. Use rock or concrete block if you can. Cedar or other untreated woods are nice but may be expensive unless you can reclaim some. Do not use railroad ties or pressure-treated wood. Railroad ties are treated with creosote, which is a very nasty coal-tar-based polycyclic aromatic hydrocarbon, which you don't want near your food. Likewise some pressure-treated wood can contain copper arsenic which is a nasty chemical that you don't want near anything you eat. With the concern about the safety of this chemical in residential applications many companies are no longer treating their lumber with it, but I would still recommend you stay clear of pressure-treated wood. Try and make the edges as natural a material as possible.

One of the advantages of a raised-bed garden is that the soil will get warmer faster because more of it is exposed to the sun. So you may want to have a raised bed to get your heat-loving plants going early. The downside of a raised bed is that it will lose moisture more quickly, both to evaporation and to excess water running out the sides during major downpours. So if you do plan on using raised beds you'll have to be more vigilant about watering them. If you're invited to a neighbor's cottage for a weekend in the hottest, driest part of the summer, make sure you find someone to water daily; otherwise you'll be very disappointed when you get back.

We love tomatoes so I plant a lot more than we need. I purchase plants from the greenhouse and start some as well, just to add variety. I try to make sure that I grow several different types of each vegetable I plant. This improves the chances that with varying conditions at least one variety will produce well. Since the greenhouse plants are invariably bigger and healthier than the ones we start indoors, they ensure that I get stuff to eat as soon as possible. If some start ripening much later that's fine; those are the ones that we'll can or freeze or pass along to our neighbors, especially the neighbor who keeps us supplied with horse manure.

I also use this technique with other plants like peppers, broccoli, cauliflower, vines like squash and watermelon, and eggplant. We start each of these indoors late in the winter as well as several times as the spring progresses, and we also buy commercial plants. This means that I'm varying the time that things will ripen so that I can try and have as steady a supply of produce as I can during the growing season.

Let's use broccoli as an example. While some well-known public figures aren't big fans of broccoli, it's an amazingly healthy food noted for its cancer-fighting properties, so we like to have a supply available from the garden all summer. So on our last frost day we put in four plants from the greenhouse. I also plant half a row of seeds that weekend and put some of our own plants, which are two or three weeks behind the commercially grown plants, in the remainder of that row. About two weeks later I'll put in another half row of plants that I started. You can do this until July. Broccoli is one of those plants that likes cool weather and can handle a frost. Invariably, despite with all my strategy. there'll be a few weeks in the summer when we have way too much broccoli, so those are the weeks we freeze a few bags and give some to the neighbors. We'll have broccoli into September and even October here, even after we've had frost. By then I'm usually pretty sick of broccoli and will take a month or so off before we start eating the frozen stuff.

You should employ a similar strategy. You'll be tempted to get out there that first weekend and put everything in and fill up the garden completely. You'll only do that once, because come July you'll have more vegetables than you can eat. The next year you'll start planting as early as you can and putting new stuff in every week until July. This will spread the season out nicely.

I find that in some years plants start better as seeds and in other years they do better as transplants. I plant my vines like muskmelon, squash, watermelon, zucchini, and pumpkin in small hills with lots of room around them to grow. I usually plant a few transplants already well under way and some seeds as well. Some years the transplants will take over and thrive. Some years they don't but the seeds pick up the slack. I like my plants to have a "Plan B."

Pests

I also do this because of the issue I have with pests. There are many small critters out there that will be thrilled that you've provided a wonderful healthy diet for them, and they'll do everything they can to eat it all themselves. The biggest problem I have is with cutworms. Cutworms look like caterpillars and they hide in the soil over the winter. In the spring they wait for green growth to eat. They will come out at night and wrap themselves around your plants and cut them off just above the soil. Then they'll pull that plant down in to the soil and munch on it during the day. It sounds pretty cute, but a couple of cutworms can take out a whole row of peas or beans before you know it. I deal with them in a variety of ways. My first strategy is to plant more seeds than I need. Then as they sprout I keep my eye on the plants and look for the telltale signs of cutworm activity, either plant stocks sitting there with their tops missing or the top pulled into the ground. If you're not closely examining the rows on a regular basis cutworms are even easier to spot because you'll look at a row and you'll see areas with lots of seedlings started and then big holes in the rows. I just head right to those areas and start carefully digging through the soil until I spot the cutworms. I use the plural because they often work in groups. Sometimes I throw them in a margarine tub with water to drown them, but usually I'm so fed up with them trashing my beautiful seedlings that I take great pleasure in just squishing them with my fingers. I usually take a "live and let live" attitude towards creatures, but I draw the line at mosquitoes and cutworms.

To protect the transplants in the garden, I take a toilet paper roll, cut it lengthwise, and then cut it in half. I then wrap half a toilet paper roll around the bottom of the plant and push it down half an inch into the ground. This usually keeps the cutworms away. I also use a sacrificial plant strategy. Using seed from last year, I start a large flat of some plant that germinates easily. Lettuce usually works well. In between each of my peppers or tomatoes I put in the two- or three-inch-high lettuce plants. Over the next few days cutworms in the vicinity will be drawn to the lettuce, which they'll slice off. It's easy to see the lettuce knocked down and easy to retrieve the culprits responsible for the work.

My other major pests are potato bugs, or the Colorado Potato Beetle. These will overwinter in the vicinity of your garden and will wait until the potatoes start sending up shoots. Then they'll fly over, start munching on the leaves, and lay their eggs on the underside of the green growth. Potato bugs are fairly easy to spot, so I always start a row of potatoes very early, even before it would be normally considered safe. This row will attract all the potato bugs in the area and make them easier to spot. Then for the next few weeks I'll keep checking this row daily and drowning any bugs I find and scraping any eggs off the underside of the leaves with my fingers. These eggs clusters are easy to spot when you pull back the leaves while standing over the plant, because they are bright orange against the green of the leaves. By planting one row early you'll find that the bulk of the potato bugs will be drawn there and you can concentrate on eliminating them. As you start other rows later you will have some bugs but they will be far fewer. By then your garden will be in full swing so you may miss some, but a plant can handle a few bugs. If one of those egg clusters does hatch in a week or two your potato plant will be full of rapidly growing orange potato bugs. They are easy to spot and you can use a badminton racket to knock them into a bucket that has an inch of water in the bottom to drown them. Stay on top of them though because if you let them they will eat all the leaves that are taking the energy from the sun and storing it in the potatoes underground.

I focus on potato bugs because I think you should put a lot of effort into your potatoes. This goes back to the food chapter, where I suggest that potatoes are the perfect food to grow in your garden, rich in energy, protein, and vitamin C. They store well, so you can eat them all winter, and the ones that you don't eat you can plant next spring and start the process all over again. Planting one potato will net you eight to ten potatoes in the fall depending on the variety you grow and what sort of summer you have. Once you buy seed potatoes you'll really never have to buy them again. Every few years I buy a few new varieties just to keep some different DNA coming into the gene pool. For several years I got too reliant on Red Pontiac potatoes that we really liked but that were susceptible to a blight. Some of the potatoes ended up with a dime-sized black spot in the middle called hollow heart. This wasn't a problem if you were cutting them up to boil or make home fries, but if you were baking them you wouldn't spot this until after you cut it open. So now I make sure we have at least three varieties. A seed catalog will list the characteristic of each variety. You'll want to get some that mature quickly so you can eat fresh-dug potatoes as soon as possible, and you'll also want some that mature more slowly and are better for storing.

For 30 years my garden has been a great experiment in seeing what works and what doesn't. I get a large sheet of stiff cardboard and I tape a large piece of paper to it and draw a map of the garden. I attach a pencil to it with a wire so I don't lose it. Then every time I plant things I note the variety. This row was "Lincoln" shell peas and this row was "Sugar Sprint" edible pod or snow peas. Then next winter when I'm sitting down to order my seeds I decide if I want to try that variety again. You can also keep your seed orders each year in a binder, but I find keeping track of things as I plant them helps me remember names. Eventually snow peas will form a pea and look like a regular pea pod, so if you're doing a stir fry and want to make sure you get the edible-pod peas, a map of what you planted is a great aid.

As the summer progresses I continue to plant new rows of seeds to keep a steady supply of vegetables ripening. Beans will keep producing for three or four weeks but eventually many of the beans will be too large and tough to enjoy, so it's nice to have some new plants that will provide you with more tender beans. As for the beans that get too big to cook up with dinner, you can leave them until the fall. At that point once the pods dry out you can pick them and dry them completely. Then during the winter as you're sitting watching TV you can open up the dried pods and remove the bean seeds. You've got a couple of options with the seeds. These beans are just like the ones that come in a can of baked beans or chili, so you can use them in cooking. You also can save to plant next year. Remember, it's always a good idea to put in more seeds than you need and thin the rows if too many germinate.

Since you're going to plant more seeds than you need, you're going to have to thin those plants as they emerge. If you look on a seed package for beans it may say that plants should be spaced 12" apart. If you saved seeds from last year and sowed them very densely you may have 10 or 20 come up in that 12" zone. It's always hard to visualize what your garden will look like in August as you see those plants emerge in May, but you have to try to because most of these plants will get very big, especially if you've been aggressive in your compost application. So ideally you won't want too many plants too close together because they'll crowd each other out. They'll compete for water and nutrients and this added competition will reduce the yield of each plant. So you are actually better to try and follow the directions on the package. I say this as the worst plant thinner on the planet. Oh, I've got my reasons not to, namely those cutworms that I never really trust to stop lopping off my seedlings. At a certain point your plants will just be too big and strong for a cutworm to gnaw through, so you have to thin. It's the hardest thing you'll have to do in the garden because it goes against every one of your plant nurturing instincts, but it has to be done. Sometimes I do it over several weeks, which makes it easier and gives me some reassurance that I won't thin to a reasonable level and then have cutworms take out the few plants I have left.

In terms of carrots I do some thinning but leave them pretty dense. I find carrots can be very finicky to germinate, so once they do germinate I hate to thin them too much. Carrots will also grow fairly compactly if you leave too many in one spot. When I thin I make sure that I wet the soil well first, which makes it easier to extract the plant you want without damaging neighboring roots too much. Since carrots are slow to germinate, by the time I'm ready to thin them they usually have a pretty good crop of weeds mixed in with them too. In fact, my thinning is probably better described as a good weeding that takes some carrot seedlings as I go.

Seed Saving

Pea plants are like beans, so if you don't get a chance to harvest them in time and the peas get too tough to enjoy you can just leave them until the pods go brown and harvest the peas to use as seeds for next year. Lettuce and spinach plants that are allowed to grow without being harvested will eventually send up a seed shoot, which will produce wonderful flowers. If you leave these long enough those flowers will brown and you'll end up with dozens of small seedpods. You can harvest these seeds and use them next year. When you cut your first broccoli the plant will continue to send out smaller shoots that will be smaller heads that you can eat as well. If you don't harvest all of these, the broccoli head will become a mass of flowers that will form seeds that you can harvest for next year once they dry.

Plants like these produce seeds every year (annually). There are other plants that produce seeds every second year (biennially). Carrots and cabbage have to grow one year and then be overwintered and planted again the following year. That second year they send up shoots that form flowers and then seeds. Depending on where you live and how cold your winters are, you can sometimes leave these vegetables in the ground over the winter and they'll start their seed cycle the following year. If you get very cold winters you can either heavily mulch them with straw or hay or leaves or you can pull them up with some root on (in the case of the cabbage), put them in a bucket of sand in your root cellar, and then plant them back in the soil the following spring. As you spend more time gardening you'll get better at seed saving; it is a financial boon, and it will also be a good way to meet others who may want to swap you for the seeds they've saved. It's just like trading baseball cards, but these trades will provide you with food!

Some seeds will germinate in your garden without your help. Each spring you'll find lots of "volunteers," seeds which fell off vegetables that you let go to seed last year and are now coming up on their own. My policy usually is to get rid of volunteers or move them where I can use them as sacrificial plants to attract cutworms. Sometimes these seeds will come from plants like tomatoes. Any tomato that ends up being left on the ground will be filled with hundreds of seeds. Sometimes these volunteers come from plants that are "hybrids," having taken characteristics from several plants. If their seeds are allowed to germinate they don't always make the best plant if left to mature. Over the years I know I have had some hybrid carrot seeds end up in my seed collection, because if I have a thick, wide row full of carrots, every tenth carrot will be white and won't taste good. I just discard this and don't worry about it because the other nine are fine. These white ones may be coming from wild Queen Anne's Lace, which is a wildflower also called "Wild Carrot" that grows around the garden and can cross-pollinate with carrot flowers. There are several excellent books on seed saving and if you find yourself enjoying the satisfaction that comes from being able to generate your own seeds each year you should read one of these books.

Watering

After compost the other key to a successful garden is water. If you're growing in a city with town water you're lucky. Many parts of North America experience water shortages every summer, which can be very hard on a vegetable garden. So as I discuss in the water chapter, you need to have as many rain barrels as you can, especially if you're serious about growing a good portion of your own food. On a smaller scale you'll be able to use watering cans to water your vegetables from your rain barrels. Try and water in the morning because it will allow the water to get to the roots without rapidly evaporating, which it will do later in the day once the sun comes out and warms the soil. If you have a large garden and limited water you may want to use mulch to help with water issues. Mulching materials are basically the same things that you're adding to your compost heap; but instead you're just going to apply them directly to the garden. So once your plants are up you can put shredded leaves, grass clippings, straw or hay, or wood shavings around the base of each plant. If you're putting in transplants you can put mulch around right away. Some gardeners like to get mulch on as early as possible to try and warm the soil quickly and inspire vigorous growth early on. You'll want to water mulched vegetables heavily to make sure the water gets through the mulch and gives the roots a good soaking. Then when the sun comes out the mulch will keep that moisture from evaporating out too quickly.

Obviously if you're putting leaves or grass on your garden and keeping it wet it's going to break down just as it would in a compost pile, and this is a good thing. You're adding that organic material to mulch the soil at the same time as you're helping to preserve some of the moisture the plants need. You can appreciate that with a large garden you're going to need a lot of mulch and a lot of inputs into your compost heap. This is where you'll have the crisis of conscience each fall: "Do I take just one more trip 20 blocks over where I know they've got bags and bags of leaves out at the curb? I'll burn gas, but I need those leaves!" Perhaps it's time to contact the city and see if they'll dump a load of leaves on your driveway rather than driving them miles away to the municipal composting facility. You can expect the initial reaction to be, "We've never done that and can't help you," which is why you'll need to contact your municipal councilor and get her working on your behalf. Many municipalities will let you go to the site to collect the compost once it's been created, so why not suggest that you're just saving all that diesel-fuel expense by having leaves delivered to your place.

As our summers get hotter many cites are also issuing watering restrictions, which is just one more reason to make sure that when it rains you collect lots of water. If you're in the midst of a drought and you know a big storm is on the way it's time to get creative. Maybe that old kiddie pool that isn't used any more can be filled up once the rain barrels are full. Vegetable plants need water and they will go into distress if they don't get enough of it. One thing you want to make sure of is that you encourage deep root growth, and you do that by watering less often but very thoroughly. In other words, rather than watering every day and only wetting the top one inch of soil, you should water every third day but put multiple cans of water on so that water gets down five or six inches or more. The deeper the roots the less susceptible the plant will be to drought damage. While you'll feel better putting a bit of water on every day you're not doing the plant any favors. Tough love means watering every few days but watering vigorously.

As your garden gets larger you may want to look into an irrigation system. A drip irrigation system can be hooked up to a raised rain barrel, a water faucet on your home, or even a solar-panel-powered pump from a nearby pond. A pressure regulator ensures that a slow and steady amount of water is gradually applied to the plants out of holes in the pipes 12 to 18" apart. This follows my deep-water recommendation by putting water on gradually and letting it spread over a large area and soak down deeply. When you water like this the soil's capillary action draws the water both horizontally and vertically. Think of dipping the corner of a big fluffy towel in water. Even if you just dip it in for a few seconds and then remove it, if you keep doing this over time the entire towel will be soaked because of the capillary action of the water moving throughout the towel. In a sandy soil the capillary action will tend to be more vertical, drawing the water down more than out. In a clay soil you'll have the opposite effect, where the water will be drawn horizontally and not as deeply. Hopefully your soil is a mixture of these types and will move water both ways.

One way to ensure that water travels well in your soil is to limit compaction by keeping the soil around the plants well tilled and making sure that your soil has lots of humus or organic material. Where does your soil get that? Compost! It all comes back to compost. Keeping your plants watered and growing them in a rich well-composted soil are the two keys to a bountiful harvest.

As the summer wears on you'll get into the groove of watering and looking for pests and enjoying your bounty. So many of the things you grow are best eaten while standing in the garden. Those first peas that you take out of the pod, rich in iron and fiber and sweetness, are a real treat and provide the greatest health benefits when eaten raw. Most of our peas never used to make it to the dinner table, so now we grow even more so we can have some with meals and freeze others for the winter. Even beans taste better eaten raw. If you grow your carrots as densely as I do and your soil is moist enough you can thin them in the summer, eat what you thin, and give its neighbors room to grow.

I grow vegetables like onions in a variety of ways. Most of them I grow from bulbs that I get through my seed catalog or the local feed mill. I transplant some that I've bought from a greenhouse, and I always throw some seeds in as well. I also grow every kind I can find including red, Spanish, and cooking onions. I grow onions close together to start, and then as they get bigger I thin every other plant to use right away. This leaves room for the remaining onions to get really big.

Garlic is a unique member of the Alliaceae family, which includes onions. It is planted in the fall. The bulbs overwinter and are one of the first things to send up shoots in the spring as soon as the snow is gone.

They grow vigorously in the spring and early summer and are ready to harvest in July. You'll know it's time when the lower leaves start to brown. After you've harvested the garlic and hung it to dry, you'll have more room in the garden to start some late beans, lettuce, and spinach.

Each year you'll get better at selecting those vegetables that do well in your garden and do well in your kitchen. There is always a tendency to grow every vegetable you can imagine and then discover that you don't end up eating a lot of them. I love turnips, once a year, at Thanksgiving. So I don't worry about starting them very early and I don't plant too many. Every year we plant kale and Brussels sprouts. These vegetables are health titans and if eaten daily would make you the healthiest person on the planet. Unfortunately I find it difficult to eat too much of them. The good news is that they keep well into the winter and they stick up through the snow in our garden, so hungry deer also enjoy some of that healthful bounty in the dead of winter when they really need it. This is a luxury you probably don't have if your garden size is restricted, so start deciding what you ate and enjoyed last year and what didn't seem to be so popular and move your garden in that direction.

More and more I focus on vegetables that store and keep well. As I discussed in Chapter 13, potatoes are the foundation of both my diet and my garden. Next come carrots, which are so rich in beta carotene and antioxidants that they are one of nature's superfoods. I put my harvested carrots into buckets and fill up the buckets with moist peat moss and they last all winter in the root cellar. When they start getting a little soft as we get closer to spring it's time for more carrot soup and carrot cake! Onions store extremely well and complement the carrots and potatoes, so I grow lots of them. I also grow a lot of garlic, which lowers your bad cholesterol and helps fight colds. These all store without any energy required.

Next come things we can freeze like tomatoes, beans, broccoli, cauliflower, and peas. These are great in soups and sauces all winter. We also grow a lot of basil, which is wonderful to add to recipes during the summer. To preserve it we chop it in a blender with olive oil and freeze it in a thin sandwich bag or ice cube tray. Then it's easy to cut a chunk off and throw it in with a soup, pizza sauce, or even scrambled eggs to give them that "upscale" look and taste. I call them "basil-infused scrambled eggs." They're incredibly pretentious and our country neighbors moan when I wax poetical about them.

We grow a lot of corn because we have the room and because I love corn. When you see the size of the corn stalks you know they take a lot out of the soil, so I make sure I move them around each year. If you have limited space it may not be the best vegetable to grow. It's best if you have 3 or 4 rows of at least 15 or 20 plants so that they will pollinate properly. When I grew corn in the city the squirrels seemed to get more corn than we did. Here in the country the raccoons stay away from our place until the corn is ready. That's when "Morgan the Wonder Dog" sleeps in our fenced garden to keep the raccoons at bay. Commercial corn can have a fair amount of pesticide on it, but fortunately we've never had to spray and still don't have major pest problems. Once we start eating our corn we always cook way too much; then with what's left over we slice off the kernels and put them in the freezer. With our Thanksgiving and Winter Solstice feasts this year we had sweet wonderful corn from our garden. An important reason to give thanks.

As the summer winds its way into fall your garden should still be providing lots of food for your table, and hopefully the weeds will have slowed down as well so that less maintenance is required. This is the time of year when some of those slower-growing vegetables like squash will be ready to eat. Since they're going to end up in the root cellar, I harvest and store the best specimens and eat the others. Any sort of scabs or blemishes on vegetables will tend to provide an entry point for the microbes that will bring about early spoilage, so we try and store only the nicest looking ones.

Now it's just a waiting game to see how long we'll go before a frost. All those beans and things you started when the garlic was harvested in July are now at their prime and you'll find yourself listening to the weather forecasts to see if there's a danger of frost. When the warning finally comes you may want to decide if there are some vegetables you particularly like that you want to keep going. In the patch where my garlic was I've usually planted beans, lettuce, spinach and a few broccoli and cauliflower plants, so I throw a big plastic tarp over it to protect it from the frost. It's important to remember to take it off the next morning because if it's sunny it can get pretty hot under the tarp. I continue fighting nature for a week or two until I finally just resign myself to the fact that nature bats last and just let it take its natural course. Some plants will survive and some won't. Before we let that last big frost hit we make sure the freezer is full of beans and tomatoes and other things we want to freeze. Some plants like broccoli, kale, Brussel's spouts, and squash will all do fine and won't be damaged by an early frost. In fact some will actually start tasting better as the cold weather enhances the natural sugars in the plant. I still may not have stored all my carrots and potatoes yet but that's fine, because they're underground and even though their tops might get nipped the root that I'm concerned with will do fine.

It's your choice whether you're going to leave the garden until the spring or start cleaning it up in the fall. If you've had any problems with tomatoes or peppers remove those plants and don't put them in the compost. If there's blight or disease it may just get back in the soil for future years. I tend to get distracted by the fall and leave much of the organic matter where it is. I'm afraid that by rototilling it in the fall I'll just leave exposed soil which is more likely to be blown away by fall and winter winds and washed away by spring rains. The one thing I do have to do is clear a section and plant my garlic. I try and do this before it gets too cold, just because it gets hard on your hands if it's too frosty out. I've always been concerned about putting those garlic bulbs into the ground just before it freezes up, but every spring they shoot right up and thrive. They are the first bits of green I see in the garden every spring and they are a welcome sight.

Every year I try new things, experiment with new varieties, and plant things differently. This year I tried planting corn in amongst the squash vines as the people native to America have done for centuries, but I didn't have much luck with it. A few years ago I tried artichokes, but they didn't mature well so I didn't try them again. Next year we're going to try planting peanuts just because they look cool to try. Every season brings failures and successes and a root cellar full of healthy organic vegetables that displace a fair amount of our grocery budget. Meanwhile we're displacing thousands of pounds of carbon dioxide that would have been generated trucking this food to us. And when the ice storm hits and we know we can't get to town for a few days, it's no problem. Our garden is still providing us with the sustenance of life. Now put down this book, pick up your shovel, and start turning over that grass in your backyard. There's a grocery store produce section out there; you just need to visualize it and make it happen!

15   Water

*Life isn't about waiting for the storm to pass, it's about learning
how to dance in the rain.*
Unknown author

*Let it rain, Let it flood these streets and wash me away
To where it makes no difference who I am*
Chapman Tracy, Let It Rain

Water is really important, yet it's amazing how many songs seem to be determined to stop the rain. Sure it can be an inconvenience, but as you get older and start to realize just how important it is, rain really is something to be celebrated.

Up to 60% of the human body is water. The brain is composed of 70% water, the lungs are nearly 90% water, and about 83% of our blood is water. Humans must replace a little more than half a gallon (2.4 liters) of water a day, some through drinking and the rest taken by the body from the foods eaten.

Water is becoming as big an issue as food and energy throughout the world. Wars are being fought over it. North Americans can be pretty wasteful with water. It's a precious resource and if you start treating it today like the precious commodity it is it'll be much easier to cope when it's not so conveniently available. So it's time you started getting in touch with water.

You're going to have 3 main water issues to deal with.

1) enough safe drinking water
2) water for sanitation such as washing and toilet flushing
3) water for growing your food

Some of the water you use you may be able to recycle. After you have a shower or bath, or wash vegetables or dishes, there's no reason you can't use that "gray water" for flushing toilets or watering gardens. As long as you use it for a "lower" purpose there's no reason not to be reusing water.

It's similar to recycling things like plastic. If you have a piece of plastic that contained food and you recycle it, it won't be reused as a "food grade" product but it could be used to make plastic wood. One of the keys to making the maximum use of a commodity as precious and essential as water is to make sure you don't waste a drop and keep putting it to use.

So let's look at each of your main water uses.

Drinking Water - Urban

Food-borne disease and illness are prevalent throughout much of the world and are one of the reasons so many North Americans have problems when traveling abroad. We simply haven't built up the necessary immunities to giardia, schistosomiasis, and so many of the other common "bugs" that lurk in water in other countries. We know that health care will not be as easily accessible in the future as governments become increasingly taxed, so not getting sick from poor quality water is going to be more important than ever.

Cities provide this amazing resource—clean drinkable water—but less than 5% of it is actually used for cooking and drinking. Of the rest, 40% goes to flushing toilets, 35% to bathing, and 20% to laundry and dishes. Then of course there are the other frivolous uses like washing cars, watering the grass to keep it looking pretty, and washing driveways—the sort of activities that remind you of a 1960s episode of *Leave It to Beaver*.

As all levels of government become taxed, it's going to be harder and harder for them to maintain the level and quality of water they provide to city dwellers. Chlorine and some of the chemicals they've used in the past will become prohibitively expensive or in short supply. Underground pipes that need to be replaced periodically will not be, and the infrastructure that provides you with your water will become increasingly problematic. So regardless of where you live you're going to need to ensure a safe and secure source of drinking water. Many places in the world experience times of the day when they have no power and water, and while it seems like a remote possibility in North America it's worth being prepared for. This means having a supply of drinking water on hand whether you live in a house or an apartment. It would be my recommendation that you use large glass containers and avoid plastic if you can. If the water is going to sit there for a while some of the chemicals in the plastic may leach into your water. Find one of those places where you can make your own wine, because they may have large glass jugs for sale.

How much drinking water you store will depend on how many are in your family, but a person requires about half a gallon of water a day. Governments are now encouraging people to stock up on water in case there are weather-related emergencies, so having water on hand is becoming commonplace. If you can't find glass containers use plastic; just get going and get it now. Every newscast showing preparations for a hurricane always shows people stocking up on water and being distraught when stores sell out. Get it now.

Drinking Water - Rural

People in the country are usually more independent when it comes to water than people in the city. They have wells that supply them with water. Hopefully you're far enough away from industrial pollution and agricultural runoff that your well provides good drinking water. If it doesn't you should look at a system to purify your water, like a reverse-osmosis unit or a water distiller. Water distilling is basically boiling water and then capturing and condensing the steam to avoid any pollutants, so it uses a lot of electricity. It's not something you could use conveniently off the grid.

Whether you live in the country or city a reverse-osmosis system will give you very good-quality drinking water. If your city is struggling with water infrastructure-related challenges it may not be a bad idea to have a system like this. You also may want to make sure you have a jug of bleach around and an eyedropper. If you can't boil water in the event of a power outage you can disinfect water with household bleach. Bleach will kill some of the disease-causing organisms but may not kill them all. But it's better than nothing. If the water is cloudy, filter it through some clean cloths or allow it to settle. Draw off some of the clear water and add 1/8 teaspoon or 8 drops of regular unscented liquid bleach for each gallon of water. Stir well and let it stand for 30 minutes before drinking it.

Your well will be either dug or drilled. If it's "dug" it means it was dug out and a concrete wall built around it. Many rural homes have a company come in and drill a well with a large drilling truck. A perforated-steel well casing is inserted which keeps material from falling back into the well while letting water trickle back in to fill it up. Inside the house there's a pressure tank which pushes the water out of the taps. After water has run for a while in the house the pressure in the tank will drop, and at a certain point the pump will click on and pump water up the well and into the tank to pressurize it to the required set point. A pump draws a fair amount of electricity. If the power goes out obviously the pump won't work. You may have a few toilet flushings' worth of pressure left in the tank but it will run out. If you have a generator to deal with power outages make sure it will handle your pump. Some smaller generators can't supply enough power when the pump comes on, and you'll hear the generator sputtering if it doesn't quit completely. So size your generator accordingly and test it before there's a power outage.

If you have your own rural water system your main concern during a power disruption is keeping the water flowing. A generator is one possibility and the other is a renewable energy system with battery backup. Again you'll want to make sure your inverter is sized to handle the pump. A pump will come on with a surge and at this point it requires a huge amount of energy. Once the water gets flowing the pump will use less and less energy over time. A good inverter will have a surge capability much larger than its rated capacity to handle this. An inexpensive inverter will not, and you'll know it because it will shut itself off when exposed to a load like this that's too large for it to handle.

Some older rural homes have cisterns, which are large concrete tanks that store water which comes from eavestroughs that divert rainwater. In drought-prone areas these are very handy, but you should monitor the water quality if you are going to be drinking it. I will always remember the story that writer Timothy Findley tells in his book *From Stone Orchard* of how badly the water smelled and tasted when they moved into their farmhouse. Further investigation discovered a raccoon that had died in their cistern. It is always a good idea to have any rural water supply tested by the local health authority.

Water Sanitation

Toilets are pretty amazing things. So are sewers. We can simply flush away the wastes that for much of human history contributed to disease and sickness. We tend to take them for granted—until they stop working. Then it's an emergency. When 50 million people in the northern U.S. and Canada were plunged into darkness in the blackout of August 2003, some learned very quickly about toilets. Most cities have backup generators to maintain water pressure, which keeps toilets flushing. Apartment buildings all need pumps because the water pressure from the city will only get the water up to the 6th floor. So people living on upper floors in apartment buildings with no backup generators or generators that didn't work had no water. I remember the television image of people walking down 20 flights of stairs to get a bucket of water to flush the toilet. Sure it was great exercise for a few hours, but if it had continued for days people would have taken to the streets. And it looked as if most people hadn't practised this on a regular basis to be in shape to handle it.

In a city you are pretty much dependent on someone else to provide your water for you. Municipalities have done an excellent job at supplying water reliably for decades, but it's important that you realize this might change soon. Anything you can do to anticipate water disruptions is going to make your life a lot easier. If the news calls for inclement weather, have a bath the night before and leave the water in the bathtub. Should you lose water pressure when the storms hit, you'll be able to flush your toilet with that water. Simply scoop out a bucket from the tub and pour it into the bowl and it will take what's in there with it. If it comes to this make sure you don't flush for each pee either. If it's yellow let it mellow. Don't flush it until you've used it two or three times.

The same can be said for your kitchen sink. Keep it filled with water. Have a couple of buckets around that you keep filled with water. If you see large plastic water jugs in the neighbor's recycling box, grab those and keep them filled up with water for emergencies. You can always use it for flushing toilets and washing hands.

Composting toilets are becoming more common in cottage and rural applications but haven't hit the mainstream in urban areas yet, and I'm not sure they will. There are two main types, some which use electricity and some which don't. Obviously for an off-grid application your preference should be the one that doesn't use electricity. Some models also use no water while others use some. If you're considering one, evaluate your priorities for the system and then get a handle on what your usage will be. You may only use it for four weeks in the summer, but if you have eight people there while you're using it you'll need to make sure the toilet can handle the demand.

From a personal hygiene perspective handwashing is probably going to be as good as it gets if you have prolonged disruption to your water supply. You can see where my bias towards rural and renewable-energy-powered living comes in. At my house my water is always available, clean and ready to drink, and it's pretty much limitless. As long as we have sun and wind we have water flowing. In ten years of living here we've never had a day without water.

People who camp are familiar with the issue of storing drinking water. If you have the large camping jugs make sure you keep them filled up. There are collapsible plastic containers with handles that you can keep at the ready. Aquatank (www.aquaflex.net) makes lightweight, portable plastic containers which can hold up to 150 gallons. If you anticipate water disruptions in your area, investing in some of these units would be a good idea.

The key today is to prepare for a water disruption and hope it never happens. I'm not advocating having a 45-gallon drum of water in the living room of your 15th floor apartment. I am suggesting, though, that you realize there is a tremendous amount of energy involved in getting the water out of your tap and that in an emergency energy can often be in short supply. Drinking water in cities is maintained with energy and chemicals to make sure it's safe to drink, and during a prolonged power outage some systems may fail. You just need to be prepared for this possibility and not be lined up with everyone else trying to buy bottled water.

Water for Gardening

Rain barrels are essential at your house, and one isn't enough. You need them on every downspout, and on the downspouts where the largest amount of water from your roof is channeled you need a few. Most of the time these are going to be for watering your garden. With the challenges of water shortages in the summer these will be crucial. In an urban environment you should keep them filled as much as you can for as long as you can. Rain barrels will give you days' worth of toilet flushing and basic water uses like dish washing should there be a disruption to your water supply. Be careful during the fall and spring, though, as freezing can crack and destroy a plastic rain barrel.

Water sitting in rain barrels will start to grow algae and get ugly after a week or two of hot weather, which is why you should cycle the water through them. Within a few days of a rainstorm use the water for the garden. Then if no rain is forecast refill the empty rain barrels from your municipal water supply. If you have two 50-gallon barrels this gives you 100 gallons of water on reserve. If they are not being filled with rainwater, after a few days drain one into the garden, then the drain other the second the day, and keep cycling them that way. In terms of gardening water this is actually better for the plants anyway because you've taken that very cold municipal or well water and allowed it to warm to the outside temperature, which will avoid shocking your plants. Allowing the water to sit will also help to dissipate some of the chlorine that is in much municipal water.

If your garden is large you'll need to get creative with water storage, especially if you live in a drought-prone area. There are lots of large rigid plastic containers you can purchase. Trying to find a source for used containers may save you some money if you can verify that they haven't had some nasty chemical in them. What about that old pool the kids don't use anymore? Or the larger pool your neighbors had a few years ago and got sick of? Maybe they'd trade it for a couple of baskets of tomatoes when your crop comes in if you can use it as a rain reservoir. That old canoe by the back fence is great for storing water when you turn it up. I'm always trying to find new ways to store water. The reality is that when you get a summer rain sometimes you'll get way more than your rain barrels can hold, so I'm forever running the extra into our old canoe and kiddy pool to increase my reserves.

The health authorities would climb the walls at my house because of all the water I have sitting around. Reservoirs of water are breeding grounds for mosquitoes which today can spread West Nile Virus. You should take action to prevent mosquitoes from breeding in your rain barrels. Commercial barrels are covered or screened. Several drops of vegetable oil is said to prevent the larvae from hatching. My suggestion is to keep cycling and replacing the water. Sometimes at my house I'm not as vigilant as I should be with my 12 rain barrels and yes, I find mosquito larvae. But my house is surrounded by ponds and the mosquitoes can breed just about anywhere around here, so it's a losing battle. Luckily the mosquitoes here don't seem to have the virus yet, but you should be careful.

The one thing I make sure of is that my rain barrels are elevated to make it convenient to get a garden hose on them. This helps with moving the water to the garden if it's a distance from the house. It also allows me to conveniently fill up the old kiddy pool and boat that I use to store the excess during downpours in the summer. I live in a very dry area and have sandy soil which dries out quickly. I also have a very large garden, so when I do get a big rain during the growing season I store as much of the water as I can.

My rain barrels are a dark plastic and the water in them will keep longer than it will in things like the kiddy pool. The water turns green pretty fast in there, so I use that water first. I don't wait too long because

I find that if I leave the soil too long after a rain the surface of the soil gets hard and will not accept water as readily. So a day or two after the rain I start using up the water that will go bad first, giving the garden a deep, thorough watering.

Water Conservation

It goes without saying is that you have to use water efficiently and not waste it. Most people know not to leave the tap running while they clean their teeth or shave. Many municipalities are mandating low-flush toilets because so much water is wasted with old designs. Up to 40% of a household's water is used just flushing the toilet. The ultimate way to avoid this is with a composting toilet. Since many people, especially urban dwellers, probably aren't anxious to embrace that level of environmental stewardship, a better strategy is to install a low-flush toilet, which uses 1.6 gallons (6 liters) per flush. An even better solution is a dual-flush toilet, which uses just 1 gallon (4 liters) for the easy stuff and 1.6 gallons for work requiring more water. In areas where water is really precious a toilet with a hand-washing basin on the back has been designed, so that the water you use to wash your hands goes into the toilet tank to be used for the next flush. It's a brilliant concept and one that is starting to recognize the importance of this precious resource. This respect for water is something that really hit home to me when we moved to the country, where we have a well and a finite amount of solar and wind energy to pump that water into our home. I keep a plastic jug beside the bathroom sink. When I'm running the water to wash my hands and I'm waiting for it to get hot, I run that cold water into the jug. During the summer I use that water in the garden and during the winter I use it on the woodstove to humidify the air.

We wash dishes in a dishpan so that we're not filling up the whole sink. I run just enough water in the bottom to cover a row of dishes and mugs. Then as I rinse the glasses after washing them, the water gradually fills up the dishpan. By the time I've finished I've used a fraction of the water of a dishwasher and none of the electricity it would use.

New products are coming on the market that also let you become much more efficient with water. With a graywater recycling system by companies like BRAC (www.bracsystems.com) you take the water that's left over from baths, showers, and laundry, filter it, and then use it to flush your toilet. It's such a brilliant concept because that's perfectly good water for toilet flushing and you've saved all the energy and expense of having your city clean and pump you fresh water for the same task.

Places that are severely water challenged, like parts of Australia, have helped develop rainwater catchment systems (www.rainharvesting.com. au). You can drink rainwater but you have to be careful. These systems start with a good gutter screen to strain out leaves and larger items from the eavestrough. Then the rainwater goes through a second debris screen before it goes into a first flush diversion. This brilliant invention lets the first rainwater that hits your roof be diverted until the rain has given the roof a good cleaning. There may be deposits resulting from air pollution and pollen in that initial water, so you divert it and use it to water the garden. When the diverted water reaches the top of the reservoir, the rainwater is then diverted into the main reservoir, which is a metal or plastic tank. The water is then filtered for drinking. When you see systems as well developed as these appearing on the market you know we have both water challenges and solutions.

Treating water like liquid gold is a responsible thing to do for the planet. Anticipating and preparing for a disruption to the water that's supplied to your home is the responsible thing to do for your family. The two goals are perfectly complementary. Once again, what's best for you is best for the planet as well.

16 Transportation

I can't drive 55
Sammy Hagar

I got a sixty-nine chevy with a 396,
Fuelie heads and a hurst on the floor
She's waiting tonight down in the parking lot,
Outside the seven-eleven store
Bruce Springsteen, "Racing In The Street"
Darkness On The Edge Of Town

Man we love our cars and we love to drive. We were all "Born to
Be Wild," to jump on our motorcycles and into our cars (and
RVs) and hit that open road. It's pervasive in our culture from books,
to music, to "road movies." Jack Kerouac's book *On the Road* is said to
have defined a generation that was mobile and always on the move. Our
whole culture is based on a one-time bestowment of fossil fuels that are
now running out.

I recently saw Matthew Simmons interviewed on the television show
The Agenda (www.tvo.org). He is the author of *Twilight in the Desert: The
Coming Saudi Oil Shock and the World Economy.* As an energy investment
banker Simmons has spent 40 years in the oil business, and early in the
new millennium he realized that we really didn't have any good data on
how much oil Saudi Arabia really had. It claims its proven reserves are
262 billion barrels, which is almost twice what second place Iran claims at
132 billion and almost nine times what Nigeria or the U.S. claim as their
reserves. But if you look at the history of the reserves OPEC countries
claim they have, you'll find they all went up dramatically in the 1980s

without any actual new discoveries. This is because within the cartel you can only pump oil based on your "proven reserves," so if you want to pump more oil and therefore make more money you have to fudge your numbers.

The theme of Matthew Simmons' book is that if Saudi Arabia with its massive reserves is past peak, then the world is past peak. That doesn't mean we run out of oil tomorrow; it just means that as we begin to move down the back side of the curve and produce less the price and supply are going to become more volatile. With the economic collapse destroying some oil demand we may actually hit a plateau for a while. But Simmons claims, as do many petroleum geologists, that the world has hit peak oil, and he believes it peaked in 2005. We have produced around 85 million barrels per day since then and even with the price skyrocketing to $147/barrel in the spring of 2008, the supply has not surpassed that rate. The demand was there, the supply was not.

The interviewer, Steve Paiken, said, "There are all these great new electric vehicles and hybrids. Won't that help us?" Matt Simmons' response was that it took Toyota five years to sell a million Prius and five years before that to develop the technology. There are currently 700 million automobiles on the planet with internal combustion engines that burn liquid hydrocarbons like gas and diesel. We still need a lot of gas for those vehicles and it won't be long until we see $500 and $600/barrel oil. European governments tax oil so high that many European drivers pay $600/barrel for fuel right now, and there is still lots of private-vehicle use. The difference is that the vehicles in Europe are much smaller and more fuel-efficient and people drive them less.[1]

This is all happening as many of China's growing middle class enter the car-ownership market and Tata Motors in India has just started shipping its "Nano," a $2,500 car designed to get people off bikes and scooters and into the driver's seat.

The reality in North America is that we will have to rejuvenate our railroads since many of us will begin taking the train more often. Trains are extremely efficient because, unlike a car or bus where you have rubber hitting asphalt, which results in great friction and therefore more fuel to move the vehicle, trains have steel wheels that roll on steel rails with very little friction or resistance, and they are very efficient. Compared to the energy required to get a jet off the ground and keep it moving, trains are simply light years more efficient. Rail travel will have the obvious advantage of reducing how much CO_2 we put into the atmosphere. Jets are particularly bad because they release CO_2 high in the atmosphere where it does much more damage than at lower altitudes.

Ultimately Matt Simmons' greatest insight in the interview is that we're all going to end up living in villages again. Even in cities people will drive less and trade much closer to home. Our economy is going to return to being more local as oil becomes prohibitively expensive and increasingly harder to come by. For those who were against the march towards an integrated world economy, it appears that peak oil is going to do what a concerted effort on the part of individuals across the planet couldn't, namely reverse the tide of globalization. The message from an individual point of view is that you are going to be traveling less and less, which impacts your decision about where you live. The advantage of a city of course is that with a high concentration of people in one place the economics of transit is much better. Peak oil is basically going to bring about the end of suburbia. Urban design which requires cars as the basis for all transportation is going to become increasingly unsustainable. Having to use your car everyday to go to work and attend social and sporting events and go shopping will become prohibitively expensive as the effects of peak oil start to show up at the gas pump.

Europeans have already reacted to this reality by building more compact cities. The smaller distances between major European cities also makes high-speed rail much more viable. For some trips it is now faster to take a train than a plane between major cities. North America has large distances to cover, but we'll have to either do it by train or stay home. We're going to have to take to the extreme the 1980s concept of cocooning—staying home with popcorn and a rented movie rather than going out on the town. We're all going to have to become homebodies. If you like the nightlife this will be a bad thing. If you like the comforts of home this will be a good thing. Regardless of which camp you're in, it's going to require a huge adjustment for a population that has lived with endless and cheap travel to deal with restrictions on its movement. This will be difficult for many people to deal with. Cheap and abundant energy has shaped generations of North Americans. It has allowed families to live at opposite ends of the country or in different countries altogether. I can't tell you how many people I've met lately with older children who live all over the planet. If families continue to live this way they are going to see each other less and less. Trips home at Christmas are going to become extremely expensive. They may get together every other Christmas at first, but as the rate of oil depletion increases they may not be able to afford to do it at all.

I've had participants at my workshops put questions to me like this: "My fiancé lives in Australia. He's close to his family and I'm close to my family in Toronto and we wonder which country to live in." Decades ago this rarely happened because travel was slow and expensive and the majority of the population couldn't afford it. Now with airfare so cheap and so many people going to school and traveling to other countries to explore and work it's inevitable that people from different locations fall in love. The problem is that these relationships are premised on the cheap and abundant energy model, and that model is about to come off the rails. There's not much I can tell them. Whichever place you choose to live you are going to risk one of you not getting to see your family very often. That will be a heartbreaking thing for everyone involved, but I believe it is the new reality. It brings me no joy to inform people of the inevitability of the end of oil, but I believe we all need to be realistic.

We are all going to be staying closer to home. Services like Skype (www.skype.com) that allow us to video chat with people over the Internet are mind-boggling in the amazing technology they provide, and they are going to make it a lot easier for people in these situations. Grandparents in a different part of the country are going to get to watch the grandkids open their gifts on Christmas morning over the Internet. Not quite the same as being there, but better than not seeing them at all. It will also help to eliminate some of those inevitable family squabbles when Uncle Billy has too much eggnog and starts dragging skeletons out of the closest.

The transportation issue is a good reason to consider living in an urban area. Hopefully governments will keep transit working to allow you to move about the city for work and shopping. Larger cities will also be hubs for rail transportation between cities. Some smaller cities may have more difficulty in this department because over time, as truck-based transportation has usurped railways, many rail lines have been abandoned and left to decay. In the rural area where I live all the villages and towns were at one time served by regular rail service. As cheap oil allowed people to own cars the rail lines were abandoned. The rail companies didn't like the liability of people getting injured on their property so many turned them over to governments and organizations for use as trails. While some are for cycling and walking many are for gas-powered activities like ATVs and snowmobiles. The unfortunate part is that as these trails are abandoned by vehicles that require gas which people can no longer afford for recreational activities it will be very difficult and extremely expensive to ever return them to rail service. The railway ties and steel rails are long gone. It requires a huge commitment on our part to resurrect them.

So transportation is going to be much more of an issue for rural dwellers. Or it's going to be an issue for someone choosing to live in a rural community and believing that they will be able to continue to live as we have for the last 50 or 75 years. It is unlikely that you'll be able to live in a small village 45 minutes from a larger city and commute to work each day. Unless you have an extremely high-paying job, the economics of long commutes is going to be increasingly prohibitive for the average North American worker. If you work in the technology field or have a job that requires just Internet access and a computer then telecommuting may work, allowing you to work from home. Perhaps you could spend most of the workweek at home and commute to the office for face-to-face meetings once a week. Many rural communities now have high-speed Internet so there's no reason you can't have face-to-face meetings with co-workers in cyberspace. Our family's Apple laptops all come with a camera installed and we can video chat with our daughters in the city two hours away whenever we want. Saves time, gas, and carbon emissions.

For a while people in outlying communities will have to form car-pool systems that allow a number of people to drive in one car to work in the city. There will be a transition period as higher oil prices inspire a more aggressive search for the remaining oil, but we've found all the giant or elephant fields and the new discoveries are always smaller and much harder to get at. During this period it will appear that with some discomfort we'll be able to maintain our happy motoring lifestyles and in the short term it may not be so bad sharing a ride to work and getting to hear about what idiots other people's bosses are too. But in the long term your goal must be to be independent and not require regular driving, anywhere. If you can scrounge up some gas, great, but plan in the future on "staying put." This gets you back to your "where to live" strategy. If it's rural, there is an advantage to being able to walk to town to the grocery store and other amenities.

One of the questions most of us have when confronted with the concept of "peak oil" is, "What's the big deal?" On the surface this is a fair question. Most of us don't understand the ability of fossil fuel to displace manual labor, and that is the big deal. Three spoonfuls of crude oil contain the same amount of energy as eight hours of human manual labor. I know this from my garden. I know how much manual hoeing and shoveling just a bit of gas in my rototiller displaces.

Every time you fill up your gas tank, you're using energy equivalent to one person's manual labor for two years. A barrel of oil equals 8.6 years of human labor. A human lifespan could produce about three barrels of oil-equivalent energy. That is discouraging! Your whole life's work distilled down to three barrels of crude oil. To truly understand the power of this amazing stuff called oil you need to do the "push your car test." Head over to the mall parking lot early in the morning or the school parking lot on a Saturday. Make sure it's relatively flat. Turn the key so that the steering and brakes work. Get someone to steer and put the car in neutral. Now get behind your car and push it. Push it for as long as you can. If you can get it moving I'll bet after about 2 minutes you'll be completely bushed. If you own a truck or mini-van you may not even be able to get it moving at all. So now that you have an appreciation for the weight of that vehicle, think about the energy required to drive around the city. It's enormous. Now think about the energy that's required to get it up to highway speed! The mind boggles. Yet we've all come to take this little miracle for granted.

I've read that the average suburbanite with an SUV in the driveway has more power at her disposal than the Pharaohs had with 1,000 slaves to build pyramids. The problem is that we just don't appreciate the potential in that magical gas we put in our tanks. The other problem is that we do not really have an alternative. There is no substitute for gasoline. There is no economical, readily available source of energy to replace the potential energy in a gallon of gas. Not the sun, the wind, not hydrogen, nothing.

We all have high hopes for the hydrogen fuel cells that we keep hearing about, but hydrogen is not an energy source—it's an energy carrier. You have to make it from another fuel, right now usually natural gas. So it doesn't offer a solution to our rapidly declining fuel reserves. What about electric vehicles powered by solar and wind? As wonderful as this sounds there simply isn't enough sun and wind power to move around the country the way we're used to. We still need to rely on the existing electrical transmission system which uses coal and nuclear power. The reality is that the grid is pretty much at capacity now so it would be very difficult for a large number of cars to switch to electric.

There are 700 million cars in the world and over 135 million cars in the U.S., so the move to electric vehicles is going to take a long time.[2] People keep their cars on the road longer these days, especially during an economic downturn, so you won't find a huge percentage of vehicles being replaced today anyway. Even if automakers come out with electric cars and plug-in hybrids, they will not make up a significant percentage of the fleet for a long time.

In the meantime people are going to continue to buy and burn gasoline and this is going to lead to the decline of world oil reserves, with gas prices going up significantly as supply falls.

We all know that driving is a bad thing for the environment, so using less gas is actually a good thing. Right now if you're considering a new car make fuel efficiency your number one priority. After oil prices dropped from $147/barrel we all quickly got used to cheaper gas prices again. Don't be fooled. This will not last. Remember that gas is going to become increasingly expensive and in some cases in short supply. So you want to make sure your car can travel as far as possible on a full tank. You'll notice I've been saying new "car" rather than "vehicle." Your days of SUV and mini-van driving are over. You need a car. A small, fuel-efficient car. If the kids are a little cramped in the back, too bad. If they don't each have their own beverage holder and DVD player they're going to have to deal with it. Get 'em in therapy. They're going to have to start realizing that they should be grateful to be moving around in this magical, independently powered steel box which is a miracle unto itself. The days of frills are over. If you're concerned about driving seven kids to a soccer tournament, get to know the local rental car agency. Rent a truck or mini-van when you need one. Don't be driving around in it 30 days a month for the 1 day a month you really need the capacity. Or let the other parents drive those days.

You should also give some thought to a diesel, especially if you have a long commute right now and you don't see reducing it in the near future. Diesel engines are on average 30% more energy efficient than an equivalent gasoline engine, and with newer, cleaner turbodiesel and common rail diesel injection motors, emissions are much improved from the old days of smoke and noise. The problem with gas or diesel fuel in the future is going to be supply. I believe this may be more acute with diesel fuel because you will be competing with uses that may be deemed more essential than your commute to the mall. Trucks which move goods around the country and tractors which farmers use to grow food use diesel fuel. If we get to the stage where there is fuel rationing, it may be harder to get diesel fuel. In the fuel crisis of the 1970s there were stories of sugar being put in the fuel tanks of diesel cars because truckers didn't like having to compete for limited diesel fuel. Luckily cars today have lockable fuel caps.

The upside of diesel fuel motors is that you can run biodiesel in them, which you can make from waste vegetable oil from restaurant deep fryers. While this sounds easy and cheap, there are a number of steps involved to make good-quality fuel. Supply of waste vegetable oil could become an issue as well. As the price of crude oil skyrocketed in the spring of 2008 restaurants that were once happy to have home brewers haul away their waste vegetable oil found industrial waste-oil haulers were prepared to pay for the privilege and in some cases bidding wars broke out to purchase the waste oil. If you're counting on waste oil you could be shut out of the game if crude oil prices rise high enough. An excellent source on biodiesel is William Kemp's *Biodiesel Basics and Beyond*. It's the most complete guide available today on how to make ASTM-quality biodiesel.

You should also consider a hybrid vehicle, which most manufacturers have available. An even better option is a plug-in hybrid. Today's hybrid cars have an electric motor to assist the gas motor, but most of the energy in the car still comes from the gas motor. A plug-in hybrid boosts the battery capacity, which allows the car to run as an electric for a bigger percentage of the time. Some people may be able to run errands and take short trips on electric power alone. The batteries are either charged by the gas motor when it's running or by the grid when you plug it in.

Some people look to motorcycles and see a motorized two-wheeled machine and assume it has fuel economy similar to that of a bike. This often isn't the case. The average fuel economy for a mid-sized motorcycle is about 60 miles per gallon (4.7l/100 km). This is worse economy than most hybrid cars and is comparable to regular gas-powered compact cars which can carry five adults and provide heating, luggage storage, crash protection, and shelter from the elements. So if fuel economy is your priority don't be looking at a motorcycle.

Scooters are becoming popular in cities but they have the same drawback as a motorcycle in that they require gas. I recently saw a Vespa gas-powered scooter advertised that had only marginally better fuel economy than my Honda Civic, although I'm sure there are models with better gas mileage if you do enough research. The Vespa looks like a blast, but I believe driving it in a city is much more dangerous than driving a four-wheeled vehicle, and without significantly better fuel economy I'm not sure I see a huge advantage to it. If you're just using it for shorter trips in town a ZENN electric car will give you four wheels on the ground and the ability to charge it with your solar panels or from the grid at off-hours when electricity is cheaper.

The overarching theme is that the days of personal car-based transportation are drawing to a close. How much money you choose to invest in a car should be tempered by the realization that ultimately you will not be able to afford or even locate liquid hydrocarbon (gas or diesel) fuel for it. If you get an electric car you need to evaluate the range of the battery system and remember that if large numbers of drivers switch to electric cars electric utilities will have to raise their rates to deal with the increased demand. While you'll be able to purchase solar panels to charge your electric car you'll have to ask whether this is the best use of this very expensive, high-quality power that you've generated. Would it not be better used powering your home rather than your car? I think you'll find that your home will take priority and that you'll have to find alternatives for transportation.

I provide the following illustration to give you an idea of how much energy our current lifestyles consume. Energy can be broken down into calories, giving us a standardized unit to measure diverse things. A gallon of gasoline for example (4 liters) contains 31,000 calories. This chart shows how much energy is required to drive 1 mile (1.6 km) to the grocery store to buy dessert for dinner tonight. I'm a huge dessert fan!

You can see that the most efficient form of transportation available to anyone on the planet is a bicycle! That's good news because it's good for the atmosphere and bikes are pretty affordable for a good percentage of the planet. A bike is almost twice as efficient as walking because it's using those amazing round tires and gearing to accomplish more work with less effort. Those tires make it much easier to move your body weight than just walking. In the case of riding or walking the calories come from food you eat that is converted to energy for your muscles. To

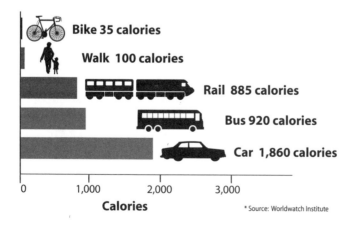

Bike 35 calories

Walk 100 calories

Rail 885 calories

Bus 920 calories

Car 1,860 calories

0   1,000   2,000   3,000

Calories

* Source: Worldwatch Institute

ride your bike to the store you need to eat about an apple and to walk to the store you need a banana. Sounds like you're going to be making a fruit salad for dessert!

Then we start using fossil fuels and the efficiency of moving you goes way down. Your share of the energy required to move that fully loaded bus or train is 9 times more than if you'd walked and 25 times more than if you'd ridden your bike. Things get really out of hand when you decide to drive your car to the store. You can see your car uses 50 times more energy than if you'd ridden your bike.

I have not included the calculation for a jet in the chart and it hardly requires mentioning that a jet uses an obscene amount of energy. While most people don't take a jet to the store, air travel continues to grow every year. If you've flown in a jet and heard the engines wind up as you take off you get a sense of the energy that is required to get that 200-ton fully loaded Boeing 767 off the ground. The mind boggles at modern flight. And once you get that sucker airborne you've got to keep putting the pedal to the metal to get up to its cruising altitude. Then you've got to keep that 200 tons of metal moving through the air, defying gravity. It is difficult to get an average number of calories for air travel because it depends on the type of jet, the distance traveled, the number of passengers on the jet and a number of other variables.

The impact of a jet engine burning kerosene or jet fuel and spewing that CO_2 and water vapor high into the atmosphere is much worse than the effect of CO_2 created on the ground. The atmosphere is much thinner and less resilient to such an intrusion. As George Monbiot says in his book *Heat: How to Keep the World from Burning*: "If you fly you destroy other people's lives." He of course is talking about those who live in low-lying areas of the world that are particularly vulnerable to the ravages of climate change, like rising sea levels in the Maldives and Bangladesh.

I hope Monbiot's statement will discourage you from flying and I hope even more that it will inspire you to get out on a bicycle right away, because bikes are very efficient. Insanely efficient! So you're going to get one. My attitude has always been to buy a cheap bike because bicycles tend to get stolen. An inexpensive bike won't have the smoothest gears or the nicest brakes, but I've always felt that if I have to work a little harder to get the thing moving it's a better workout for me, and I'm way ahead in terms of the energy required to walk the same distance. The three main bike types are mountain bikes, racing or road bikes, and hybrids, which are a cross between the two. Mountain bikes have the big knobby tires and are good if you're going to be riding over very rough terrain like trails or streets that are in really rough shape. City roads can be pretty broken up and with sewer grates and streetcar and train tracks a wide tire is considered best.

Road bikes have very thin tires and are good if you plan on riding a really long distance. The thinner tires mean less resistance and therefore less effort to travel a given distance. The downside to those thin tires is that you are much more likely to take a spill when you hit a pothole or broken pavement. Some are thin enough that they will actually slip between the grates of a sewer cover, although most cities have been changing these over to diagonal ones that are safer for bikes. So you'll be able to ride farther on a racing or road bike but you'll have to be more careful about obstacles on the road.

A hybrid, as you can imagine, is a cross between the two. It will have intermediate tires that are thinner than mountain-bike tires but wider than the tires on a road bike. They are better for a long commute in a city with marginal or questionable roads. Many of the newer designs also follow the much older design in terms of having a bigger seat and higher handle bars so you don't have to ride crouched over, which can be hard on your back over a long distance. I like my road bike for long rides and don't mind being stooped forward because it means my body is more aerodynamic and I don't have the same drag and wind resistance. Michelle loves her hybrid which has a seat built for a woman's body and higher and more comfortable handlebars.

I'm happy with a $150 bike and if it gets stolen it's not the end of the world. Our latest purchase was an electric bike and it is the coolest thing ever! Electric bikes have been refined over a number of years and their technology is getting much better. The original electric bikes used lead-acid batteries that were very heavy and had limited range. Our new Schwinn bike, which cost $1,000, uses lithium polymer batteries. The range is 16 miles (25 km) and the battery should take more charges and last much longer than a lead acid. This bike has three modes: you can just ride it yourself, have it assist you as you peddle, or have the DC motor take over and do all the work, like a motorcycle or moped. These bikes are great because they let you get some exercise but they also help extend your range with the motor. If you like to pedal, an electric bike helps you maintain a much higher speed so that you're more likely to ride than drive. I find it amazing going up hills because you continue to pedal and even though you lose a little speed you can keep up a good head of steam on a fairly steep hill. It's completely brilliant!

If you were to use an electric bike to commute to work and your workplace has no shower facilities you might want to use the motor on the ride in so you don't get too sweaty and then use mostly pedal power on the way home to relieve the stress of the day. When you get home from a ride you can charge the battery by plugging it into a regular outlet. Since we live off the grid the electricity comes mostly from our solar panels, so we have a solar-powered bike! How cool is that! In fact this would be a perfect way to get you into solar power. Put up a panel and small inverter and you can use it to charge your bike. This is where solar-powered transportation becomes a reality. You remember from your "Push the car in the parking lot" test just how heavy your vehicle is and just how much energy your engine needs to get that big mass of steel moving. Energy consumption in vehicles in large part comes down to weight. A bike is very light, so using solar power to help with transportation becomes a reality if we're talking about something like electric bikes.

If I was living on the electricity grid, the electricity to get me into town and back would cost me about five cents. Not bad. Sure it's not optimal on a snowy day, but for a good chunk of the year an electric bike is an excellent alternative to a car.

My neighbor who rides a motorcycle asked about the noise of an electric bike. I assured him an electric motor is virtually silent. Then I realized he expected it to roar like a Harley. The best I could offer was to attach a baseball card to the front fender with a clothes pin and letting it slap against the spokes as you ride. It won't attract biker chicks, but it will attract enviro chicks. Was this an appropriate thing for a feminist to say? I think not.

There are many electric scooters coming onto the market and again you could use solar power to charge these. One thing to remember with some scooters is that in certain jurisdictions they are considered a vehicle, like a car, and therefore you need to follow standard licensing and safety regulations. Some scooters try and get around this by including pedals that allow you to pedal in a worst-case scenario, but they are heavy and the pedals are an afterthought, often very spread apart and uncomfortable so that in reality you couldn't ride it very far.

One final possibility for transportation is horses. I can hear you groan and say, "We're not going back to *Little House on the Prairie* times I hope." Well, no one is going to force anyone to wear crinolines and bonnets while feeding the pigs, but if you live in the country riding a horse can make a lot of sense. In the rural community where I live you occasionally see someone who has ridden a horse into town for the novelty of it. There are lots of horses throughout the country right now and many of them spend sedentary lives loitering in horse barns and paddocks, waiting for their owners to ride them on the weekend, when it's warm, and there are no bugs, and there's no chance of rain, and there are no sales on in town.... so we might as well put them to use. They love being needed and used and if you're far enough out why not ride the horse in? As the price of fuel rises I have no doubt that the most creative retailers in small towns will set up horse "parking spots" with water and hay and those hitching rails you see in all the old westerns. It's a chicken and egg thing right now. Most towns don't have places to comfortably tie up your horse while you're in town. So if you build it, they just might come.

Now the purest environmentalists out there will argue that a bike is still a better way to get yourself into town because those wheels and gears are so much more efficient than four legs carrying a 1,500 pound horse. This may be true, but as a gardener who doesn't ride horses I can tell you I love horses. It's the end result of all that hay munching that I love for the garden. In an area like ours with marginal soils, hay is one crop that grows exceptionally well. So turning sunshine into hay and then having the horse turn that hay into manure to put on my garden makes perfect sense to me. I need to replenish my soil, and while I use some green manures there's nothing like composted horse manure.

Your challenge if you decide to use a horse for some of your transportation needs will be feeding it. Even though you don't have to worry about purchasing increasingly expensive liquid hydrocarbons like gas and diesel, you'll still need fuel. Hay will require you to have either the money to purchase it from a farmer who grows it or the acreage to produce it yourself. You would be surprised at just how much land you need to feed a horse. Depending on how much you're working it a horse might require five to ten acres of hay to support it during the year. As the price of food increases farmers are going to turn some hay into commercial crops for human consumption, which will raise hay prices. If you have enough land to support horses great, but you may be similarly tempted to turn that land into crops for human consumption. There are so many variables in this calculation that I can't attempt to make a recommendation on whether or not to keep horses. Many people love horses and if you're putting them to work all the better. If your land is poor, improving it with their manure is an excellent idea. For centuries horses displaced huge amounts of human labor, until fossil fuels became available and displaced horses. Looks like we're going full circle and the term "horsepower" will no longer refer to the 400 horses under the hood of an overpowered sports car but to the amount of energy that big, beautiful creature out in your paddock can provide.

One thing that is certain is that the future of transportation won't look like the present. I grew up watching *The Jetsons* and assumed I'd be zipping around in a cool little space car by now. While we humans may one day possess this technology it's a long way off, and in the meantime our predominant form of transportation relies on fossil-fuel-burning internal combustion engines that have a limited future. The earlier you start preparing for this the better. The sooner you decide to make sure you don't require a long commute to work and commerce the better. If you are going to drive, use the absolutely most fuel-efficient car you can find. As a backup plan buy some bikes. Buy a mountain bike for off-road and a road bike for longer trips. Start getting in shape. Start exercising those long-dormant muscles that you don't use when you step on a gas pedal but come in handy for pedaling a bike and moving your mass a distance on the most efficient form of transportation there is. Then once you've got yourself back in shape, reward yourself with an electric bike to increase your range and displace your car miles more and more. Make its purchase part of your "Green Energy Plan" so that once you get your solar electric panels installed charging up your electric bike will be one of their first uses.

The winds of change in our endless happy-motoring lifestyles are blowing hard. They could very quickly turn into a whirling tornado of chaos for people who are dependent on cars for their livelihood. Now that you know the big wind is coming you can go down to your storm cellar and emerge in your spandex, gel-cushioned cycling shorts with your new bike held high over your head like the environmental superhero you are! Better for the coming challenging times, better for your health, and better for the health of the planet!

17 Health Care

He who enjoys good health is rich, though he knows it not.
Italian proverb

To keep the body in good health is a duty...otherwise we shall not be able to keep our mind strong and clear.
Buddha

I was born in Mechanicsburg
My Daddy worked for Pontiac 'til he got hurt
Now he's on disability
And I got his old job in the factory
Warren Zevon, "The Factory", Sentimental Hygiene

Good health is something that's very easily taken for granted. Sometimes you'll read about or see a TV show or movie about someone with a diagnosis of terminal cancer and you'll vow to yourself, "You know, tomorrow I'm going to get up and be grateful that I'm healthy!" And often the next day you do, but then the next day you have that meeting with a stressful coworker on your mind, and the day after that you wake up with a headache because you drank too much coffee yesterday, and before you know it you're back in the rut of "just getting through every day."

At 50 I have been blessed with incredible good fortune and am still healthy and active. My family is as well. I feel I am even more blessed to live in Canada because of our universal health care. Every Canadian has free health care regardless of income or location. Decades ago a politician named Tommy Douglas saw the economic hardship a major illness could have on a family and decided to make national health care his mission. He fought for it for many years, through debilitating strikes by doctors and constant criticism by the right and the stigma of being labeled a socialist. But the concept grew and grew in popularity in the country until the major governing parties had no choice but to enact legislation to create a national health care system. Now someone living in poverty who gets sick gets the same care as someone living in a mansion. There is no charge for most common medical tests and procedures.

The beauty of Tommy Douglas' vision was that Canada, like the United States, was a very rich country and we could afford to take some of that wealth and spread it around so that everyone got to share in it. A universal health care system works like an insurance company in that it assumes that some people will use the system less than others but still contribute to it by being productive members of the economy; these people will subsidize those less fortunate who get sick and have to use the system. Insurance companies stay in business because lots of people keep paying premiums but don't die, so that when someone does die and claim his share, the pot is big enough to handle it.

Canadians love their universal health care just as much as they love their hockey. In fact in a recent national TV show which polled the country on "The Greatest Canadian," Canadians picked Tommy Douglas over Wayne Gretzky and other nation-building figures. Canadians define themselves through their national health care program. The system is not perfect. It has some problems and sometimes lets people down. With an aging population it faces real challenges on how it can continue to offer quality care to Canadians as they get older and require its services even more. One of the problems with the system is that Canadians don't really have any idea what it costs to operate. We don't get bills when we visit the doctor. We don't get bills when we have surgery. The hospital that provided the surgery knows what it cost but unfortunately doesn't share that information with the end user. If we were to get a statement, even though we didn't have to pay it at least we'd know how lucky we are to live in a country with universal health care. It is funded by the taxes we pay and the wealth we create working in the economy.

Americans are well aware of the cost of health care and the cost of health insurance. It's one of the reasons people are hesitant to leave employers who provide health care, which distorts the job market by reducing mobility. The U.S. health care system provides exceptional service to those who can afford it and abysmal or no service to those who can't afford the insurance. In his documentary *Sicko*, Michael Moore exposes some of the problems with the system in its present form. 50 million Americans are underinsured or have no insurance at all. A major illness can bankrupt a family. Many Americans are already used to living without health care. As people get laid off and companies go bankrupt, many more people are joining the ranks of the uninsured.

As much as I hate to admit it, Canadians will soon find themselves in a similar situation. Our health care system costs the government about $160 billion a year or close to $5,000 per person per year. A huge chunk of the money the government spends goes to health care. Now we have an economic crisis which is forcing the government to spend money to stimulate the economy and pay more to the unemployed and workers in need of retraining. The result of this economic crisis will be that governments have much less money to spend. They will have to ramp up spending to help citizens deal with skyrocketing oil and home heating costs resulting from peak oil and peak natural gas as well as rapidly rising food costs and other climate-change problems; something's going to have to give. Governments have only so much money to spend and can only go so far into debt. Sooner or later their creditors will cut them off and they will have to get their fiscal house in order. First the level of service available to health care consumers will be reduced. This will accelerate as more and more services and procedures are no longer covered. Eventually it will be obvious to most that the system is unsustainable and we'll return to a pay-as-you-go system like the one we had before we had a formal health care insurance plan.

It gives me no joy to see this inevitability. If I were running a small business in the U.S. and making very little money it's unlikely I'd be able to afford health care insurance. My daughter's two surgeries on her ear could possibly have bankrupted us, or at least severely affected our ability to pay off our mortgage and move to our rural location. I can't tell you how fortunate I feel to have had that health insurance available. So I grieve its potential loss.

The one thing I don't grieve is the abuse of the system, which basically rewards you for being unhealthy. There really is no financial incentive to stay healthy when you have health care insurance like this, because someone else will pay for your mistakes. Universal health care inspires people to abuse the system. Spend any time in a hospital emergency department and you'll realize that lots of people shouldn't be there. They have a cold. There's nothing a doctor can do for a cold. They have the flu. Sure, that's a good idea, head over to the hospital and spread those germs around knowing full well there's nothing physicians can give you.

Got high blood pressure? Why spend $75 on your own blood pressure tester when you can go into your doctor's office and have her check it for free? Oh sure, you're wasting her time, which is extremely valuable, but if it's free, why not? I have always advocated user fees with subsidies for those who can't afford them, but the Canadian model of universality means nobody pays. This system is going to be so universal it collapses under its own weight.

In the U.S. health care is going to become so expensive for many Americans that they'll just have to opt out. So suddenly there will be large numbers of North Americans no longer covered by health insurance. And what should you do with this in mind? You should get yourself healthy! You should get healthy and you should stay healthy. I realize that some illness may be inevitable and there are some things you simply can't prevent. But the vast majority of health issues from heart disease to diabetes to cancer can be prevented. You just have to be proactive about it.

Much of this book has already laid the groundwork for you to improve your health. That huge garden you're putting in is going to require lots of physical effort to till and weed and water. It's going to require you to expend lots of calories growing your own food. That food you grow is going to be mostly vegetables and they're going to taste better than anything you've ever eaten from a store. Some nights you're going to sit down to a feast of fresh vegetables and they're going to be so tasty that you're not going to slather them in butter and coat them in salt. You simply won't need to because they'll taste out of this world. Coming directly from your garden to your dinner plate, they'll be full of the micronutrients and enzymes that commercial food lacks. What your vegetables will lack is traces of pesticides and insecticides and herbicides and fungicides and all the other chemicals that don't help your body in any way. They help the farmer compete in a food system gone mad, but they don't aid in your good health.

The other thing that plate full of new potatoes and steamed broccoli and fresh corn will not need is an anchor. It won't need that big slab of animal protein that North Americans build their meals around. There's nothing better than a plate full of fresh, tasty vegetables to distract you from the saturated fat that you don't need and that in fact leads to many negative health effects. As I discussed in Chapter 13, there is growing consensus that the developed world must consume less animal protein. Cycling plants through animals to produce animal protein for human consumption is not sustainable on a mass scale. At the same time there is a growing consensus in the health community that the large-scale consumption of animal protein is having a detrimental effect on the health of those living in developed countries. There is also concern as more and more developing countries move towards a North American diet. Other countries are emulating our love affair with fast food. Of all the things the developing world should not follow our lead on, eating junk food is at the top of the list.

We have known for a long time that a diet high in saturated fat is not healthy for us. We also know that many cancers and other health problems like diabetes are caused by poor diet. One of the best sources for information on this topic is the Physicians Committee for Responsible Medicine (PCRM), an organization of doctors who focus on the causes of poor health rather than just correcting the problems when they occur. For many years these doctors have advocated a vegetarian diet because of the overwhelming data which supports this choice.

In 1948 in Framingham, Massachusetts a study was begun by William Castelli, M.D., to see what influences heart disease. The study showed that there is a cholesterol level below which coronary artery disease does not occur. People with cholesterol levels less than 150 milligrams per deciliter (mg/dl) have the lowest risk of heart disease. The study, which has been ongoing for 50 years, shows that people who live in the developing world typically have cholesterol levels below 140 and they do not develop heart disease. One hundred million Americans have cholesterol levels over 200 and for most people with coronary artery disease the level is 225.

How do you lower your cholesterol level? Stop eating it. Where do you find cholesterol? In animal products: red meat, fish, eggs, poultry, milk, yogurt, and cheese. The cholesterol is mainly in the lean portion of the meat. Chicken contains the same amount of cholesterol as red meat and most shellfish are high in cholesterol. No foods from plants contain cholesterol. So one of the best ways to lower your cholesterol level is a vegetarian diet. The other way is to increase the consumption of fiber, of which vegetables and grains have lots and animal products have none.

The World Health Organization has determined that dietary factors account for at least 30% of all cancers in Western countries. Large studies in Europe have shown that vegetarians are about 40% less likely to develop cancer than meat eaters. [1]

In the U.S. researchers studied Seventh-Day Adventists, of whom half are vegetarian and half eat some meat. The studies showed that those who avoid meat reduce their risk of cancer and heart disease significantly. [2]

Studies at Harvard have shown that people who eat red meat daily have three times the risk of colon cancer than those who rarely eat meat. The American Institute for Cancer Research (AICR) looked at major studies on food and cancer and concluded that consumption of red meat increased the risk of cancers of the breast, prostate, kidney, and pancreas.[3]

There are many reasons for this connection between meat consumption and cancer risk. Meat has no fiber and nutrients that would have a protective effect. A number of nasty chemical compounds are formed when you eat meat. I've seen ads in the paper by the local health unit explaining how to cook meats properly on the barbeque. They suggest that you wrap the meat in aluminum foil. Yea right, then you'll lose all the good smoky flavor and burnt fat. Well there is method to their madness. The fat dripping on the grill creates smoke which contains polycyclic aromatic hydrocarbons (PAH) that coat the meat. One of those compounds is benzopyrene, one of the most carcinogenic compounds in cigarette smoke. It has been known to cause cancer since the 1930s because it's the compound in coal tar that gave so many chimney sweeps their cancer. The health unit never explained why to wrap your meat in aluminum foil though. I guess they didn't want to offend anyone.

So I figured if I ate meat that wasn't barbequed I'd be okay. But it turns out that no matter how you cook meat nasty chemicals occur. Grilling, frying, and even oven broiling will create heterocyclic amines (HCA), a family of mutagenic compounds that cause cancer. It doesn't seem to matter how you cook it, bad things appear. But, you say, what about the cavemen grilling their freshly caught prey over an open fire? It's in our DNA to eat meat! I'm not sure there's any anthropological evidence to prove this, but I don't think your average caveman lived as long as we do today. Some saber-toothed tiger was bound to do him in at a young age before all those PAHs and HCAs got to him. If you want to live longer and healthier and reduce your need for health care you have to seriously look at how much meat you eat.

The same holds true for diabetes. The more fat in your diet the harder it is for insulin to get glucose into the cells. Studies have shown that diabetics who increase their consumption of grains, legumes, fruits, and vegetables while reducing their consumption of meats, high-fat dairy products, and oils are able to reduce their medications after 26 days on a near-vegetarian diet and exercise program.[4]

It doesn't seem to matter how you look at it, humans were not meant to eat animal products, certainly not in the volume consumed by North Americans. We think about the benefits of "adding" grains and vegetables to our diets to counteract the negative effects of all that saturated fat and cholesterol rather than just not eating the cholesterol to begin with and structuring our diets around plant products.

It's really important that readers of this book not feel I have a hidden animal-welfare agenda in my message. I don't. There is no doubt that eating a vegetarian diet is a more humane way to eat, especially as factory farms have forced much of our food to be raised in inhumane ways. But I also have no bones to pick with farmers who raise animals. I will remind you that farmers grow those grains and vegetables and fruits I'm advocating. As a vegetarian you're not trying to put farmers out of business, you're simply changing the end product you purchase from them. The reality is that in an increasingly food-challenged future, farmers are finally going to be able to earn more money growing corn and grain and products for human consumption than cycling them through animals to try and increase the value added and hence income they can generate.

I do not eat animals but I do eat some animal products in limited quantities. I eat eggs several times a week that are produced by my local organic farmer. I can walk over and see the chickens pecking away at the ground in the sunshine as they're meant to. I do eat some cheese and put milk in my tea and cream in my coffee. I find not being a vegan (no animal products of any kind including dairy and eggs) is a little bit easier. Having a bit of dairy increases the variety of stuff I can have when I'm at a restaurant. It makes it easier for friends if we go there for a meal too.

Of all the things that I recommend you do to improve your health and minimize the support you need from health care professionals, moving towards a plant-based diet is probably the key. So many of us have an image of farmers decades ago eating huge amounts of animal protein to give them the stamina to do the exhausting levels of work required on a farm. Today, for most of us, including farmers, fossil fuels have displaced that human effort. The problem is that many of us still continue to eat as if we were working in the fields all day. If you spend the day doing manual labor and burn 4,000 calories a day then by all means eat whatever you want and as much of it as you want. But most of us don't put in that kind of effort. And while we should be basically limiting our calorie intact to 2,000 many of us eat far more than this. It's calories in calories out. If you eat 2,000 calories a day and burn 2,000 calories a day you're going to maintain your weight. If you eat 3,000 calories a day and burn 1,500, you've got a problem. And for so many of us the extent of our exercise is the walk from the house to the car, from the car to the office, from the office to the restaurant for lunch, and then the same in the reverse.

The concept of restricting calories is pretty basic to most diets, but it's not one that I want to emphasize. One of the keys to helping us get healthy, though, is an awareness that many of us have been conditioned to larger serving sizes. Over the last few decades the volume of food that is served to us at restaurants has steadily increased. Of course this has happened as we have been doing increasingly less calorie burning thanks to our longer commutes and generally busy lives. So you should try and start eating less of everything, and one of the best ways to do this is to replace your regular-sized 10" dinner plates with smaller 7" or 8" dinner plates. This sounds pretty basic, but the visual connection between your mind and body when it comes to food is very complex. Putting a smaller volume of food on a large plate looks to your mind like deprivation. Putting that same volume of food on a smaller plate, which makes the plate look full, has a much better effect on your visual clues about being full and satiated. I know it sounds too easy, but give it a try. Find a second-hand store and buy some intermediate-sized dinner plates. Put a reduced serving on the large plate and see how "wanting" it leaves you feeling; then put exactly the same volume on the smaller plate. You will notice a difference. And then eat that smaller serving slowly and enjoy every bite. When your plate is clean, don't go back for seconds. Or if you must, go back for more beans and cauliflower and skip the grilled marinated teriyaki tofu, which is going to have more calories.

Low-fat plant-based diets work over the long term to help keep your weight down by focusing on fiber-rich complex carbohydrates, which make you feel full without eating so much fat. In his book *Eat More, Weigh Less*, Dr. Dean Ornish says: "When you go from a high fat to a low fat diet, even if you eat the same amount of food, you consume fewer calories without feeling hungry and deprived. Also, because the food is high in fiber, you get full before you consume too many calories. You can eat whenever you're hungry and still lose weight."

I do not need to list the stats on the general health of North Americans. Or the number of overweight or obese people. Not only is there a huge diet and weight-loss industry, there are TV shows focused on losing weight. It's an epidemic. In fact I'm thinking about taking this book and republishing it under the name *Cam's Guide to Weight Loss and Optimal Health*. The key is getting more exercise and moving towards a calorie-limited plant-based diet. I add "calorie-limited" because you can still eat way too much on a plant-based diet. Heck, a meal of pop and chips would count as "plant-based" since the pop is mostly high-fructose corn syrup, and a slab of black forest cake for dessert would qualify too. That's not what I'm talking about. I'm talking about the move to a diet based on starches— rice, pasta, potatoes—with lots of vegetables on the side.

The beauty of starting a move to this diet now is that there are so many wonderful "transition" foods. Not only are there exceptional veggie burgers and veggie dogs, there are plant-based products that mimic chicken, whether it be fillets or nuggets or strips or fingers. And there are products like veggie bacon and veggie sausages as well as ground round, which is a low-fat soy-based version of ground beef. These products are going to help you as you migrate away from animal protein. They are going to help you deal with the inevitable cravings you'd get if you just suddenly sat down to a plate of vegetables. Doing that is like most diets. They don't work because you have to give stuff up. Eventually you fall off the bandwagon and go back to your old way of eating. By transitioning to these types of products you won't feel as though you've given something up. I am very confident that if you experiment and prepare them just as you prepare your meat-based versions, by barbequeing or adding sauces and marinades, you'll have a much easier time of it. You can also try using plant-based protein like tofu, which is made from soybeans, as a meat substitute. The key is to prepare it with sauces and spices and attention to make it live up to its potential as an excellent transition food and protein substitute as you evolve your eating.

Many prepared products are plant-based, using things like wheat and soy protein, so you are reducing your intake of saturated fat and reducing your impact on the planet. Good for you, good for the health of the planet. And as you start to eat this way you'll find some nights you're happy to just have pasta without the meat or "ground round" in the sauce. Maybe it's a primavera night. Another night you may find that the teriyaki stir-fry you make to go on the rice doesn't need the veggie chicken strips after all. The vegetables and rice fill you up but don't leave you feeling weighted down. And then when you start harvesting those vegetables and having those plates full of fresh vegetables you'll honestly not miss "the anchor." The potatoes and the corn and beans fill you up. The weight loss is a fringe benefit. Now that you're not eating cholesterol, which you can only get from animal products, you don't need to spend your hard-earned money on cholesterol-lowering drugs. Now that you're eating fewer calories and less saturated fat and getting all that exercise on your bike and in the garden maybe your blood sugar will be more stable and you can cut back on your diabetes medication. As Dr. Neil Barnard of PCRM notes, maybe you can get off them altogether.

Now that you're aware of the fact that the world has hit peak oil and that we are on the downward side of the curve of world oil production, you're going to seriously consider your mode of transportation. Since you now have the bike you're going to start using it for more trips. Once you install the carrier on the back, you can use the bike for those quick trips to the store for small purchases. You're also going to be walking more. As the price of gas goes up you're going to start walking to the bus stop. The bus takes you to the train and once you get downtown it's a ten-minute walk to the office. Getting your body in shape like this is going to help you weather the wrenching shock of price spikes in oil that will become the norm. Yes, you may still own a car for a while, but if you're not reliant on it then these price shocks aren't as jarring. Maybe this year you'll take the train home for the holidays rather than driving.

Some people who end up in a rural location may begin heating with wood and actually cut it themselves from their property. Country wisdom dictates that "wood warms you twice." First when you cut it, then when you burn it. My experience has been that wood warms me about nine times. First when I cut it in lengths. Then when I drag it with a sled to where I can get to it with the truck in the spring. Then when I load it on the truck. Then when I unload it and put it on the sawhorse to "buck" or cut into woodstove-sized lengths with the solar-powered electric chainsaw. Then when I split it into stove-sized pieces. Then when I pile the firewood to dry in the summer. Then when I move it into the woodshed at the end of the summer. Then when I carry it into the house in the winter and put it in the wood box. Then when I take it from the wood box to the wood stove. Perhaps I overanalyze, and I am certainly not an "efficiency expert" because there are probably some steps I could cut out.

But I choose not to. I love heating with wood and I love cutting it. I sweat at just about every one of the steps and sweating means I'm getting exercise and burning calories. And that's a good thing for my health. And after a day of firewood cutting I have a slightly bigger piece of black forest cake. Using the calories in calories out equation, I can pig out because I burned it off!

It always amazes me how healthy and active many of my older neighbors are who heat with wood and have continued to garden and engage in activities that require physical exertion. Then I look at so many urban older people who have got to the stage where they basically don't do anything. They are overweight and unhealthy. Their focus seems to be when their next trip to the doctor is and what medications they are on. This is absolute madness. It focuses on poor health and a belief that chemicals are what keep you healthy. This is false. What keeps you healthy is what goes into your body for energy and using your body constantly to do useful work which keeps your mind challenged and makes you more positive and more likely to ignore a little muscle ache and get out in the garden and hoe some weeds.

There is no doubt that as we age we can't accomplish as much and start slowing down. But that doesn't mean we should park ourselves in the garage and throw away the keys. It means we need to work smarter. It means maybe we pay that young kid down the street to do some of the heaviest lifting in the garden and we take on a little more of the "management" work. We all like to joke about the common image of 85-year-olds from Sweden spending the day cross-country skiing, but there is some truth to this. In his book *The Blue Zone: Lessons for Living Longer From the People Who've Lived the Longest*, author Dan Buettner researches the places people live the longest throughout the planet. His first observation is that the people who live the longest eat a plant-based diet. They're not necessarily vegetarian, but meat is an occasional thing rather than the default. This is good news for you if you really would miss that turkey on the holidays. While there are lots of excellent substitutes, you don't have to give it up completely.

Buettner also finds that the people who live the longest often don't have access to all the accoutrements that make our lives so easy. All those machines and appliances are actually making us less healthy. This all comes back to my "Three Benefits Theory." Using a push lawn mower is better for you, better for the environment because you're not burning gas, and, since you don't have to buy that gas (or the high blood pressure medication), better for your personal independence. Lots of the people in places like Costa Rica where there is high longevity don't have cars; they use bicycles and walk.

The more you start acting as if we've run out of oil, the better shape you'll be in and the less you'll need our health care system. It will also help you mentally deal with the inevitable while you still have the option of using the system when you have to. If you have chosen to live across the country and away from your immediate family and a parent gets sick, it's important to ask yourself what you would do if you couldn't hop on a plane and fly home. This time you have the option. Maybe next time a super spike in oil prices will put a flight home out of your price range. Then you may need to evaluate where you've chosen to live. If family is important to you then migrating closer to home is something you should consider soon, while you can still afford things like a cross-country moving company. And family is important. Studies show that another key to mental health and well-being is a sense of belonging, to a family or community. Sometimes this is a church, sometimes it's a sports team, but a sense of community is crucial to good health. This way you're not in it alone. You have a support network. For some people that takes the form of their family, but it can take many other forms.

If you don't belong to a network or community now some of the recommendations I've made in this book may help you find one. Let's say you live in an apartment in a city, which can sometimes be pretty isolating. Well, you know you need to start a garden, so track one down. Many will be community gardens which are coordinated by volunteers. There will be a board or committee that meets regularly to deal with issues that come up like dealing with the property owner or neighbors and security in terms of making sure that members' food is staying in their plots. So you should join the board. It's amazing how a common cause like this will bring people together. Some of the meetings may be held at local eateries. You might take on the job of surveying members of the garden, which will involve meeting and chatting with them while they work in their garden plots. You'll be surprised at how quickly you'll find people with similar interests.

I'm not suggesting you go out with the goal of forcing yourself on other people in the hope that they'll accept you. I've been trying this for years and ended up in the bush four miles from the nearest human being. But volunteerism has numerous benefits and finding yourself part of a community is one of them. It's good for your health and an excellent way to share information about dealing with the changes that are happening.

One of the benefits of you growing your own food is discovering how much better-tasting food is that is grown without the chemicals that occur in pesticides and other toxic compounds. This may lead you on a journey to discover just how many potentially harmful chemicals you encounter on a daily basis and how many of them you can eliminate. Often there are simple and natural alternatives that are far superior for your health. With others, like plug-in scent dispensers, removing them from your home will save electricity and eliminate chemicals that the products pump into your air. As you start going down the list of how many of these products you don't need you'll find you can expose your-self to a much lighter load of potentially hazardous chemicals. When you discover that the reason potatoes look so nice in the store so many months after they are harvested is that they are sprayed with a disinfectant you would hesitate to use in your house, let alone eat, you'll find that organic potatoes look more appealing. Better yet, you'll discover that the ones you grew that may have a few scabs on their skin and don't look as if they'll win any awards at the local agricultural fair taste just as good if not better and reduce your exposure to something that would not be considered a health promoter.

The move away from chemicals and toxins in your home is going to lead to better health. I remember watching an interview with three experts at a cancer conference. Two spoke about the progress that was being made in curing cancers. The third was obviously the odd man out. He questioned the war on cancer stating we shouldn't be putting so many resources into curing people once they get cancer, we should be trying to prevent them from getting it in the first place. He suggested, "We know what causes cancers. It's our diet. It's smoking. It's cosmetics and the toxic chemicals women use to color their hair." Ever since then I have watched numerous fundraisers for various cancers and I see all those women running to find a cure for breast cancer and notice how many of them are blond, and I assume that many of them are coloring their hair without realizing what they're doing.

So it's time for you to start being your own health care control board and analyzing what you put in and on your body. The information is usually out there and easy to find on the Internet. Try and track down sources that are as impartial as possible, but don't assume that the information you get is perfect. Sometimes what's best to default to is just how close to nature the product is and how many chemicals have had to be modified to accomplish the task. Whether there is a link between hair dyes and cancer is not the point. If the person applying the product to your hair is wearing gloves and using foil wraps to keep the chemicals off your scalp, trust your instinct that this may not be a health-promoting activity. Never assume because the government allows a product to be sold that it's healthy and non-toxic.

If you have a pool you should be aware of the aromatic hydrocarbon C_{12}, or chlorine. Downloading an MSDS (Material Safety Data Sheet) will give you some understanding of why there is a skull and crossbones "Toxic" warning and "Corrosive" and "Dangerous to the Environment" warnings on the container. The toxicology section on the sheets suggests, "Toxic by inhalation, ingestion and through skin contact. Inhalation can cause serious lung damage and may be fatal. 1000 ppm (0.1%) is likely to be fatal after a few deep breaths, and half that concentration fatal after a few minutes." Under the Legislative Footnotes on another sheet I found this: "[1]Ingredient listed on SARA Section 313 List of Toxic Chemicals. [2]Ingredient listed on the Pennsylvania Hazardous Substances List. [3]Ingredient listed on the California listing of Chemicals Known to the State to Cause Cancer or Reproductive Toxicity." Yuck! The stuff sounds nasty! And I'm pouring it in my pool and hot tub and my kids are swallowing it? I'm not suggesting you can eliminate all hazardous substances from your life, but in an age where staying as healthy as you can is going to become increasingly important, maybe having chlorine around isn't the best idea. This is a decision you'll have to make. And if, as the MSDS says, there is a possible association between chlorine exposure and certain types of cancer then are there alternatives? You need to start being proactive about your exposure to things that may harm your body and start avoiding them or finding alternatives. Would a salt-water pool system work?

Don't municipalities use chlorine to process my drinking water? Yes they do. So you should be using a filter to remove that chlorine. This is the sort of research you should be doing in your life to find out if processes are inherently safe. What about fluoride? Many municipalities are no longer fluoridating their water because some research has shown an elevated risk of some cancers after ingesting fluoride in drinking water. From your perspective, regardless of what your municipality is doing do you think it's safe? Could you or your family be getting too much fluoride? It's not only in your water, it's in our toothpaste; it's even in tea. There have been incidents of people in North America with fluoridated water who brush with a fluoride toothpaste and drink tea having toxic levels of fluoride in their bodies.

So you need to take a proactive approach to your health to ensure that you minimize your need for the people in the white lab coats.

Comparing this book to others in the field, you'll find that many other books deal with first aid. Some of the more "extreme" think it's important that you know how to deal with a gunshot wound. But I'm not going down that road. You should start building a medical library and good first aid books should be part of it, as should books on herbal remedies and natural healing techniques. If you develop an ailment after losing your job and health insurance, wouldn't it be nice to find that brewing a tea from a plant you can grow in your garden can help relieve the symptoms? Our home is surrounded by poison ivy, which periodically one of us comes into contact with. Another plant that grows well here is jewelweed or touch-me-not, which is said to be a natural way to relieve the discomfort of poison ivy. Michelle has a tendency to get poison ivy in the spring when she's planting the garden, but jewelweed doesn't flower until the summer. So last summer she picked and froze some of the jewelweed so she could use it in the spring!

There is a whole world of natural health out there that doesn't require you to send your hard-earned dollars to pharmaceutical companies. You'll need do some reading and some research, but it will be worth the effort. Illness is often just something being out of balance in your body, and you can often find a way to bring your body back into balance without the trauma of a prescription. When I see prescription drug ads on TV that say, "Side effects may include dry mouth, heart murmurs, night sweats, vomiting and diarrhea, temporary drop in blood pressure…," I start wondering if the problem you're trying to cure doesn't sound way better than the side effects of the drug.

Herbal and homeopathic cures may not work and you may still need to consult a doctor, but if the symptom doesn't seem life-threatening I believe it's worth a try. Or just waiting to see if your own immune system can handle it. Remember, that new diet of fresh fruits and vegetables and your new exercise regime of walking to the store and riding your bike to work have made you much healthier and your immune system much more robust. So give it a day or two and see how your natural defense mechanisms work before you call out the big guns.

If you don't already you should have a first aid kit around. It's always a good idea to take a CPR course. While your family will be much less likely to need CPR with your new low-fat diet and exercise regime, you may have visitors who need it. If you live in a rural area it might not be a bad idea to have something like Benadryl around in case someone has a reaction to a bee sting. Benadryl, which you can buy over the counter, is an antihistamine which can help if someone is having a mild reaction to an insect bite or sting. In extreme reactions you may want to have an EpiPen, which is an autoinjector of epinephrine (or adrenaline) for use when someone goes into anaphylactic shock. While you would hope to never have to use this, in a worst-case scenario it would be good to have one around. The pens cost about $100, are good for two years, and should be replaced after that time.

If you live close to a hospital emergency room an EpiPen may not be that necessary, but for people in rural areas they are a good idea. I cannot tell you the number of times I've rolled over old logs or hay bales and disturbed a hornet's nest. I had about four of them descend on my ear once and man did it hurt. Anaphylaxis can occur if you've been stung a number of times in the past. Your body's immune system becomes sensitized to the allergen, for example, bee sting toxin, and even though you've been fine in the past you may suddenly react. Allergies to peanuts and drugs like penicillin can also bring on these events. If the ambulance is a good drive from your place, consider an EpiPen.

While you can't be prepared for every possible health emergency some supplies and books will help. You can always run to the Internet but in an emergency it may be difficult to find the most reliable source of information quickly. There are many excellent websites which can help you deal with some of your health issues. In fact some doctors are finding they are consulting with patients who already have a good idea of what's wrong and what the solution is. If a doctor prescribes a certain drug the response is, "Well wouldn't XYZ be a better option?" We're entering a new age in health care. Many of us are starting to realize that while doctors have spent many years gaining knowledge it often has to be broad-based and they therefore may not have expertise in the problem you're having. So you have to become proactive. This is good conditioning for the days when you can't afford a doctor or the health care network is just not there to serve you so well.

This isn't necessarily a bad thing. Humans have always relied on natural remedies for health problems. They have also relied on healers who focused on less invasive procedures. Childbirth has always been incredibly dangerous for women, but midwives for centuries used techniques of massage and strategies to move babies that didn't want to make their entrance into the world. It's only recently that the North American medical establishment has begun to recognize what a valuable contribution midwives can make to the delivery process. A woman who has worked with a midwife and delivered her baby at home is already on the road to realizing that we don't always need the high priests of the medical profession. They are wonderful to have and in an emergency those of us that can afford medical care are extremely lucky. But the system is starting to strain under the weight of a bloated bureaucracy and a population that's getting older and fatter and less healthy. Your best bet is stay out of the system as much as you can.

18   Safety and Security

Early and provident fear is the mother of safety.
Edmund Burke, 1729-1797, political writer, statesman

After a shooting spree, they always want to take the guns away from the people who didn't do it. I sure as hell wouldn't want to live in a society where the only people allowed guns are the police and the military.
William S. Burroughs, 1914-1997, writer

I'd love to do a character with a wife, a nice little house, a couple of kids, a dog, maybe a bit of singing, and no guns and no killing, but nobody offers me those kind of parts.
Christopher Walken, actor

I know what you're thinking. This chapter is a no-brainer. Safety and security? Get a gun! Well all right, yes and no, but at least let's look at some options. Lots of people already have guns, but I get a sense that a lot of people reading this book won't own a gun. They haven't had to and they don't like the idea of us all arming ourselves. And a lot of the people who have already heavily armed themselves are never going to read this book anyway. And clearly, lots of Americans have firearms.

It's estimated that there are 200 million firearms in the United States. More than 50% of Americans say they own guns. That's a lot of guns. And the numbers are climbing quickly. In the fall of 2008 when it looked likely that Obama would win the election, gun sales went up significantly as some Americans felt a democratic president would be more likely to restrict gun sales. But they have continued since. A recent article in *The Post-Star*[1] talks about how local gun dealers are experiencing increased sales in every class of gun from handguns to shotguns and in ammunition sales as well.

According to *The Orlando Sentinel,* "Selling bullets may be the most secure job in Florida as long as supplies last. After months of heavy buying, gun dealers across the state are experiencing shortages."[2]

The following is an excerpt from an editorial in *The Wall Street Journal:*

> Gun sales continue up. The FBI's criminal background check system showed a 23% increase in February 09 over the previous year, a 29% increase in January, a 24% increase in December and a 42% increase in November, when a record 1.5 million background checks were performed. Yes, people fear President Obama will take away the guns he thinks they cling to, but a likely equal contributor to what *The Wall Street Journal's MarketWatch* called a 'gun-buying binge' is captured in the slogan on one firearms maker's Web site: 'Smith & Wesson stands for protection.' People are scared.
>
> They are taking cash out of the bank in preparation for a long-haul bad time. A friend in Florida told me the local bank was out of hundred-dollar bills on Wednesday because a man had come in the day before and withdrawn $90,000. Five weeks ago, when I asked a Wall Street titan what one should do to be safe in the future, he took me aback with the concreteness of his advice, and its bottom-line nature. Everyone should try to own a house, he said, no matter how big or small, but it has to have some land, on which you should learn how to grow things.[3]

For years I have been giving my Thriving During Challenging Times workshops and have divided them up into "soft landing" and "hard landing" scenarios. The soft landing scenario suggests that although we do have many challenges governments, businesses, and individuals will be able to make the changes necessary to deal with these multiple converging problems. The "hard landing" scenario has a more urgent plan of action and it's the one people don't like to think about too much. When I suggest that if governments can't deal with all the challenges, and if businesses don't adapt fast enough, we're all going to be dealing with a radically different reality, my recommended course of action is therefore much more aggressive. One of the things I discuss is personal security and it's one of the things people want to think about the least.

It's not something we want to think about, but unfortunately if you take all the problems I've discussed in the first part of this book to their logical conclusion, there's a distinct possibility that our lives will in fact change radically and our neighborhoods won't be as safe as they are now. I can remember when I first became aware of this possibility. It was when I read James Howard Kunstler's book *The Long Emergency*. In it he goes through the peak oil scenarios and concludes that the world will be a more dangerous place in the future. I can remember his recommendations on where to live in the United States and his feeling that the south might not be one of the better places. He suggests that there are still some racial tensions which may resurface with a bad economy and that people in the south have a bit of a rebel attitude that may not make for the most tame evolution to the new reality of a fossil-fuel constrained world. He also says that many people in the south are armed and that these elements all make it a good place to avoid if you can during The Long Emergency.[4]

I remember thinking that he was a bit too extreme for me and that human nature is better than he allows in the scenarios he describes. Then along came Hurricane Katrina in August 2005. At first it seemed as if the city had been spared, but then the levees started breaking. It quickly turned into a humanitarian disaster. It was difficult to watch the poor souls stranded at the Superdome and realize that this was happening in a developed country, the richest on the planet. I remember watching the news and hearing reporters say they could hear shots as people on the ground fired on helicopters. That didn't make any sense to me. Why would people be shooting at someone trying to help them? It was on the Internet that I was able to solve the puzzle the mainstream media had created. Apparently the people at the Superdome were predominantly black and helicopters outside were rescuing some white people; when black people discovered the selectiveness of the rescue they let their feelings be known to the pilots of the helicopters. These sorts of stories were confirmed in many non-traditional sources and included stories of blacks being prevented from getting across bridges out of the city to more white-dominated neighboring areas.

While I couldn't understand this happening, it really did convince me that James Kunstler was onto something.

Since the election of Barack Obama the official government attitude towards these challenges has changed as well. Director of National Intelligence Dennis Blair went before the Senate Select Committee on Intelligence in February of 2009 suggesting that the failing global economy is a bigger threat to U.S. security than al-Qaeda or the spread of weapons of mass destruction. He stated that the economy could trigger a return to the "violent extremism" that we experienced in the 1920s and 1930s.[5]

While the official unemployment rate continues to sound manageable at 9.5% this number has been heavily politicized over the years, with populations of workers omitted to make things look better than they really are. If you include people no longer looking for work because they are discouraged and part-time workers who cannot find full-time employment, the unemployment number is closer to 17%.[6]

While the 500,000 jobs the U.S. economy shed in December 2008 was brutal, in January 2009 it lost another 655,000 and in February 2009 another 650,000 people lost their jobs. Throw in another 2 million Americans in prison and you're counting an awful lot of unemployed and unhappy people. Some of these people will not be able to find gainful employment with the current state of the economy and will be forced to try and take things from others.

The U.S. Army War College recently released a report called *Known Unknowns: Unconventional 'Strategic Shocks' In Defense Strategy Development* which suggests that the U.S. military must be ready for:

> ...violent, strategic dislocation inside the United States where widespread civil violence would force the defense establishment to reorient priorities to defend basic domestic order and human security. Under the most extreme circumstances, this might include use of military force against hostile groups inside the United States. Further, Department of Defense would be, by necessity, an essential enabling hub for the continuity of political authority in a multi-state or nationwide civil conflict or disturbance.[7]

Countries all over the world have already experienced low-level instability because of the worldwide economic crisis so it only makes sense that the military must consider it a possibility here. While none of us likes the idea of the military being called and marshal law being declared, the challenges governments face today are so severe that members of the military and intelligence communities are now discussing it in public. They're concerned, so you should be concerned too. I'm not suggesting that you be concerned to the point of immobility; I'm merely suggesting that regardless of how loathe you are to think of such a possibility you need to start considering how to protect yourself.

A good place to start is with your neighbors. Groups like Neighborhood Watch are a good starting point to get to know your neighbors and share concerns about security issues in your community. A sense of community may be one of the strongest weapons you have to feel safer wherever you live because you'll know people are looking out for one another. It's one of the advantages of intentional communities and places where multiple families choose to live communally. But if that's too extreme for you, your neighbors will be key to your personal security. It's time to fight your shyness and get to know your neighbors. It's time to have a block party. This year get everyone to bring their fireworks to the middle of the street and have one big display. Or have a street garage sale. There's no better way to meet your neighbors than by buying their old roller blades at a garage sale.

Another good idea to increase your security is to acquire a dog. If you're a little dog person, the unfortunate reality is the bigger the dog the better. Maybe outfitting your Chihuahua with a little voice modulator that makes his bark deeper and louder will help. Well, I'm not sure that exists. Even just having a sign that says "Beware of dog" can help. We live in the country and have a medium-sized dog, but he barks vigorously at strangers' cars. It always surprises me how intimidated people are by our dog because he is a sweet, lovable dog and I doubt he would hurt a fly. But I've seen couriers refuse to get out of their trucks until I called him away, so it seems to work.

Country properties sometimes have long laneways, so a remote sensor that warns you when someone is coming down the driveway is a good idea. It just removes the element of surprise. There's nothing worse than being engrossed in a book or the garden and suddenly having someone standing behind you.

Or you could have a home alarm system installed. These range from inexpensive window warning devices to full-scale remotely monitored systems. Obviously the more functional your alarm, the more expensive it will be. And if you are going to have a security firm tied into your system you have a higher purchase price and the ongoing cost of support. I can't make a recommendation; this will depend on where you live and how secure you feel your neighborhood is. But the reality is that there are desperate people out there and desperate people do things they may not otherwise do. You have to decide if you're comfortable with someone breaking into your home. It's not a good feeling to come home to find that someone has gone through your stuff and taken things you worked hard for. After experiencing a break-in a few years ago we took steps to avoid a repeat in the future.

You may wish to keep more cash on hand and actually have some precious metals in your home in case you can't get to it during a "bank holiday." You're used to having valuables like jewelry around, and now you're about to up the anti. It may be time to have an alarm installed and working and well publicized on your windows to reduce the likelihood of a break-in, regardless of how well you've hidden the silver coins in the old paint cans in the garage. You may even like the idea of an alarm system for when you're in the house in the event of a break-in. Much of this will depend on your income and your comfort level.

Your comfort level will be elevated if you have some self-defense training and now may be a good time to do this. You may not have time or be inclined to put in the effort to become a black belt, and if the intruder you encounter has a gun, it may not make any difference anyway. But some training will give you basic self-defense strategies that will help you be more at ease. One of the key benefits may be the level of confidence it helps you exude in stressful situations. Research shows that criminals will pick victims based on how they walk and what sort of vibe they give off in terms of their self-esteem. If they see someone in a parking lot looking unsure and nervous they are much more likely to target them than the person strutting across that same parking lot as if they owned the place. If you start walking with a "You want a piece of me? Then bring it on" swagger, you are much less likely to end up the victim. Criminals don't want a confrontation. They want the upper hand and if looks like you're itching for a fight they'll be more likely to wait until the next person strolls by.

These are all excellent strategies to deal with your and your family's security. You must start thinking about them whether you like it or not. And one thing you must consider is a gun. There, I said it. You might not have wanted to hear it, but I've got to discuss it. I grew up shooting a BB gun and a .22-caliber rifle, but once I started living on my own in the city I had nothing to do with guns. In fact standing behind a cop in line at the local coffee shop and looking at her handgun used to freak me out.

Then I moved to the country. We feel very safe where we are but we are three miles from our nearest neighbor. We had several incidents with people arriving at our house whom we weren't particularly comfortable with. One evening a car pulled into the driveway and the driver told me that a truck was on fire near our place. His cell phone didn't work so I called the fire department and then went down to investigate. When I got there the truck was pulled off the road and completely engulfed in flames. If there was someone in the cab it was too late and I frankly was not comfortable being there alone. I don't know if you've come upon a truck on fire on a deserted rural road, but there was nothing in my upbringing that made me particularly well-suited for this.

Later it was determined that the truck had been torched for insurance, and that was about the time I bought a gun. One of the police officers I spoke to said, "You know you're about half an hour from the detachment in the south, and half an hour from my detachment in the north, and when you call in from here and you're right in the middle it's sort of a toss-up as to which one of us responds." His message was pretty clear to me. Whereas in the city it would take three or four minutes for a cop to show up at my place if I called 911, here it would probably be half an hour, so for half an hour I'd be on my own. If all the officers were on calls elsewhere, it could be longer.

I consider myself somewhat of a pacifist and do not hunt. But I also feel an obligation to protect my family, and in this location a gun is simply the best option. I took my firearms course and got my license and bought a shotgun. The beauty of a shotgun is twofold. First, it sprays a wide pattern of shot, which means you don't have to be deadly accurate. A friend told me that in the prison where he worked the guards in the towers were armed with rifles. They rarely needed them but when there was a disturbance they were usually fairly stressed, their hearts racing, and they were not accurate in their shots. So their rifles were replaced with shotguns because as long as you shoot in the right direction there's a fairly good chance you'll hit something. When you confront someone and you have a shotgun, they know the drill.

The second benefit of a pump-action shotgun is the sound it makes as you move a shell from the magazine into the chamber. You can load three to five shells into the magazine, waiting to be used. When you're ready to shoot you pull on the wooden slide or fore-end and it takes one shell into the chamber ready to be fired. Doing that makes the very distinctive "che che" used-shell-out, new-shell-in sound. You know the sound; you've heard it in movies. Arnold is constantly doing it in *The Terminator* movies. It is a sound that intimidates. It says, "The person at the top of the stairs has something that will do me a lot of harm if I choose to go up them." And with a shotgun, after that first shot you can quickly pull the next shell into the chamber ready to fire. The person at the bottom of the stairs basically has to ask himself, "Do I feel lucky today?"

When you make that distinctive cocking sound with the shotgun, it's virtually impossible to tell if a shell has indeed been loaded, so if you have someone in your house who refuses to fire a weapon, they can still cock a shotgun and point it at an intruder and that intruder still has to ask the same question, "Do I want to risk the consequences of their pulling the trigger, or do I want to leave quickly?"

While I still do not hunt I am becoming much more comfortable with firearms and I would suggest that if you do purchase one you become very comfortable with it. There is no sense owning a weapon and fumbling with it when you need it. Holding a weapon is a very scary thing for me. I do not like the feeling of owning a machine that could easily take someone else's life. Heck, if I don't handle it properly it can take my life, so I'd better know what I'm doing.

A shotgun can be a very intimidating thing for anyone. I have a 12-gauge shotgun which just about knocks me over when I fire it. I have to brace myself as if I'm standing in a hurricane-force wind to stay balanced. A 20-gauge shotgun would probably be a better option for a woman if you're thinking of a long gun. The other option is a handgun. These are banned in Canada but can be purchased and owned in the U.S. I will not debate the merits of handguns being legal or not. I will simply say that if you are in a state that allows handguns and you feel your security is not as sound as you might like, there are some people who feel a handgun offers a solution. The same holds true as for a rifle: join a gun club or shooting range and use the gun regularly. You don't have the luxury of a wide coverage area with a handgun. You fire a bullet and it has to be aimed accurately to have the desired effect. If you have the gun know how to use it, be confident that you'll be deadly accurate when you do, and hope and pray you never need it.

If you own a gun make sure it is properly stored in a secure location that meets local legal requirements. You may have to store the ammunition separately from the gun. I know that my legal storage requirements mean that I need some warning if I'm going to use it. We recently had a black bear in the backyard. It was wonderful to see and luckily the dog was inside and didn't notice it, which gave Michelle and me time to admire it. If someone were being attacked, though, and I had to use the gun in self-defense, I would hope that the bear was just waking up from hibernation and was slow and dozy, because I'd need a few minutes before I was back with the gun to scare it off.

There are lots of other more novelty ideas for security. One is that you should marry into a big family. That means more strong backs for the fields and more security if things get a little out of hand. You could always build an earth-sheltered home and make sure the front faces away from the road so no one notices you're there. You could buy 5,000 acres in Montana and have the house in the middle of the property with a nice big "buffer zone" all around.

The best security I think comes from living in a "community," a place where people identify with the community and help each other out. There was a recent discussion on the *Life After the Oil Crash* website (www.lifeaftertheoilcrash.net) about living in the bush with a gun. The logic breaks down if four people confront you. Many people suggested you would therefore be better off living in a community of 30 people. And of course the response was "Yes, 30 people with guns."

I think we will ultimately be returning to a more local economy where we know the people we trade with very well. This local economy will exist in our community and it will form tighter, safer bonds. While I think it's human nature to want to help out other people less fortunate than ourselves, people will be expected to work and contribute if they are able. Just taking from someone else won't be acceptable in this local economy. I believe we will return to a more local economy where members who trade in the economy will create a much more livable society and a much safer place for all its members.

If in the meantime you haven't found the community, get in shape, get a dog, get an alarm, or get a gun.

19 Money

Money, get away
Get a good job with good pay and you're O.K.
Money, it's a gas
Grab that cash with both hands and make a stash
New car, caviar, four star daydream,
Think I'll buy me a football team
Pink Floyd, "Money"

I try to do the right thing with money. Save a dollar here and
there, clip some coupons. Buy ten gold chains instead of 20.
Four summer homes instead of eight.
LL Cool J

Writing a chapter on financial independence during the most severe economic crisis of our lifetimes takes on great significance. You want to offer some omnipotent secret to financial success, some tip or strategy that will allow readers to profit from the collapse and retire to the beach when it's all said and done. If you're looking for a magic bullet or a miracle drug that is going to sort out this financial mess, you've come to the wrong place. There simply isn't one. If there were one single piece of advice I could give it would be to lower your expectations. Man, what a bummer that concept is. I know.

It wasn't supposed to be like this. It was supposed to just keep going along the way it always had, things getting better with each generation. Kids would live better than their parents, and so on. It's not working out that way though. We are in the midst of an historic, dramatic, jarring, unsettling shift like we've never experienced before. Many economic indicators are suggesting that we are in a depression and it will be a long and protracted one. That means that the solution to your financial challenges won't come easily.

The reason it's going to take a long time to correct this situation is that we put off the inevitable for so long. For many decades people who managed large forested areas had a goal of fire suppression. They believed that as soon as a forest fire started it had to be extinguished quickly before it spread. On the surface this approach made sense, but what it overlooked was that during each year with no fire more combustible material built up on the forest floor. Over decades all those dead trees and leaves and branches and pine needles got deeper and thicker. When a fire did actually hit it was much harder to control because there was so much fuel for the fire. In fact now some forest managers actually set controlled fires to try and eliminate or minimize this problem.

A forest fire is a terrible thing, but it's part of the natural cycle of destruction and rebirth. Dead and old material is burned off. Seeds in the soil germinate, some only after intense heat, new growth begins, and the forest regenerates itself. The business cycle is like a forest, and forest fires are like the recessions or economic downturns that are an inevitable part of business. What's happened over the last decades though is that those in charge of the economy have adopted an out-of-date 100-year-old forest management model as their guide. Alan Greenspan (the U.S. Federal Reserve Chair) in particular felt it was his duty to avoid any economic hardship and keep the good times rolling. When there were indicators that the economy was about to go into a natural cycle of destruction or recession he used every means at his disposal to prevent or minimize it and get the economy growing again. So what we're experiencing today is the result of 20 or 30 years of forest-fire suppression unleashed on the economy, and the wildfire that's now burning through the economy with all that combustible material is going to burn very hot for a very long time. The destruction will be far greater than if we'd just let the smaller fires burn themselves out earlier. We're paying the price for an extended period of good economic times that were artificially created.

What came from the prolonged good economic times was that consumers started behaving in inappropriate ways in the context of standard economic theory. They began spending as if there would never be a rainy day and assuming that things like house prices would always go up. These things are not realistic and the inevitable downturn has caught many of us by surprise. But economic cycles, like nature, have a way of leveling the playing field and getting even. If you build your house in a flood plain, sooner or later it's going to be surrounded by water. Going into this downturn North Americans had little or no savings. In fact many had negative savings, owing more than their net worth.

The good news is that many of us are back to saving. We realize now that rainy days do come and that we'd better prepare. And that's the theme of this chapter. It's time to change your behavior and realize that things are not going back to the way they were any time soon so you need to reduce your expectations.

I am going to take a different approach than that of many of the most popular financial advisors you see in the media today. I'm going to suggest a radically different approach. I watch and read advice to people on how to deal with their financial problems and while the goal seems to be to get people back to living within their means there is still the assumption that soon things will straighten themselves out and we can all get back on the spend-and-consume bandwagon. I think we need a much more radical approach. Either those advisors are not being honest or their presumptions are incorrect. Things may not ever be going back to the way there were. Remember these same advisors are the ones who didn't give you a heads-up on what was coming. I don't think they believed it was possible for things to get so bad so fast.

Let me assure you that there have been many voices predicting this economic mess for many years. They predicted its inevitability and they predicted its severity. And now they're predicting its longevity and they're raising the possibility that even after a long period of adjustment things may never return to the way they were. That is something I sincerely believe and my advice is going to be based on that. I simply don't think you can trust prognosticators who were caught completely off guard by the severity of this economic mess. It wasn't on their radar, so I think you have to ignore any advice they're giving about how or when things are going to recover and return to normal. I don't think it's going to happen. The new normal is going to be one of much less economic activity and much less money to go around. So you need to start preparing mentally and financially for this new reality. And if I've got it wrong and things do get back on track, so much the better. What I'm recommending won't be detrimental to your financial health in any way. In fact, if the economy starts rolling along again as it has in the past, you'll just be in a much better position to enjoy some of the benefits that you didn't enjoy when you were so cash-strapped.

One of the best ways to prepare for an extended economic downturn is to get your financial house in order, and for most of us that means paying off debt. And our single biggest debt is usually our mortgage. During the housing bubble in the early part of the new millennium, North Americans bought bigger and more expensive homes and increased their mortgages accordingly. In a rising market this is a sound financial strategy. Unfortunately many North Americans made the assumption the market would continue going up forever, and its sudden and dramatic reversal has created serious consequences for many. As of February 2009, 1 in 5 or about 12 million Americans are underwater on their mortgages, meaning they owe more on their mortgages than their homes are worth.

That's the bad news. The good news is that the majority of mortgage holders are not underwater and have an opportunity to take a huge step towards financial independence by whittling away at that debt. You need to pay off your mortgage. Soon. You need to walk away from the financial wisdom of the day which says taking 25 or 30 years to pay your mortgage is a good strategy. It's not. You need to pay off your mortgage and you need to do it as soon as you can. The best way to keep the wolf away from your door is to get that mortgage off your back. Being mortgage-free opens up a world of opportunity. Suddenly decisions that involve a reasonable degree of risk don't result in your losing your home.

I understand what your response will be. Sure, it's easy for you to just say "pay off your mortgage" as if it's easy, but it's not. It's a lot of money and doing it early is going to be a real challenge. I understand that. It is a lot of money and it is going to be a challenge. But you're going to have to change your behavior and do it. You're going to have to focus on this goal like a laser scope on a target and commit everything you can to doing it. You're going to have to alter your behavior from being a consumer to being a saver. You're going to have to stay out of malls and stores. You're going to have to start making do with what you have, and when you really need something you're going to have to find places to borrow it or buy it used. You have to make paying off your mortgage a holy crusade that takes over your life. You're going to be so focused that you start picking up all those pennies and dimes you used to walk past on the sidewalk and putting them towards your mortgage. Americans take great pride in their ability to pull together and attack a challenge, whether it's winning World War II or putting a man on the moon. You have to make that same commitment to paying off your mortgage.

For the last 15 years I've been mortgage-free and that allowed us to uproot our family from suburbia where we had an electronic publishing business and move it three hours away to the woods. It also allowed us to evolve the business from doing work for corporate customers to publishing our own books about renewable energy and sustainability. It's unlikely we would have had the confidence to undertake such a journey if we hadn't been free from the constraints of a mortgage.

We also made a decision to purchase only a rural property we could pay cash for. We were not going to take on another mortgage after paying our old one off. It's too enabling a feeling to be free of the obligations of a mortgage to ever go back to those days.

To pay off your mortgage you have to structure it correctly and then you need to be committed to the task. Twenty years ago it was time for us to renegotiate our mortgage. Our existing bank showed us what the new monthly payments would be. A second bank did the same. Then I went to meet with the local trust company. They showed us what the monthly payments would be. Then the representative asked if we'd be interested in paying weekly. Our monthly payments were going to be about $800. I suggested that we did not have the financial wherewithal to pay $800 each week. She replied that the net amount would be the same, $800 per month, but that $200 payments would be taken out of our account weekly. Then she swung her computer screen around and showed us that if we paid weekly we would save approximately $35,000 over the 20-year life of our $66,000 mortgage.

I was blown away! First at how much money I would save, and second at how even though my current bank offered weekly mortgage payments, they hadn't mentioned them to me. Why would they if they would lose a huge chunk of profit? They really didn't have my best interests at heart, but it is a capitalist system after all.

I asked the customer rep why everyone wouldn't pay their mortgage this way? I assumed that most people weren't aware that this was an option, but she said that most people's finances didn't allow for it. They were paid monthly or biweekly and their budgeting didn't allow weekly withdrawals for the mortgage. This is bad financial planning on a number of levels. Mostly it's bad because people live so close to the edge in terms of money in their accounts. This is the wrong way to manage your money. You need to have a slush fund amount in your account that allows you to take advantage of opportunities like this.

So **Rule #1** in terms of money is to pay down debt. To do that, ensure you structure your mortgage to allow the greatest flexibility and fastest payback with the minimum of interest. This can include making the term shorter rather than longer. As the housing market was peaking and financial institutions needed more cannon fodder, they began offering longer-term mortgages of up to 30 years and beyond. This made it easier for some people who couldn't afford a home to suddenly own one. This is too long a term for a mortgage. Make your term shorter to force you to pay it down faster.

Shortly after we'd switched our mortgage to the trust company that was willing to let us know about weekly payments, we also looked at some savings we had. The Canadian government at the time gave checks to parents called a "Baby Bonus," which was designed to ensure that children had food and clothes and basic necessities. Under the concept of "universality," every Canadian child received these payments, regardless of the parents' income. Because we were frugal we were able to bank these checks. We called the account the "Kids'" "College Fund" account. Eventually it grew to about $9,900.

This was about the time we were starting to feel a desire to move out of the suburbs, and we knew the key to our financial independence would be paying off our mortgage. So we took the $9,900 and put it towards the principal of our mortgage. Our new mortgage had a feature where once a year on the anniversary date of the mortgage we could pay off up to 15% of the principal. So that $9,900 reduced our principal from about $65,000 to $55,000. That was a pretty great feeling, but it was tempered by the realization that this would probably be the only time we could do this.

As the year went on though, we decided to give it a shot again. We took a percentage of every one of our paychecks and put it into our "Five-Year Plan — Pay Off the Mortgage" account. This had previously been the "College Fund." We scrimped and we saved and we put off buying things. We wanted to take the kids to Disneyworld, but that was $3,000 we could put towards the mortgage. We needed a new bed, but flipping the mattress over would have to do for another year. The rust had eaten away at the floor under the driver's side of the car to the point where you could see the road whisking by, sort of like a Flintstone-mobile, but a new car could wait.

Lo and behold, the next year we were able to do it again, and we put $9,900 towards the principal on the anniversary date of the mortgage. I should point out that we were by no means well off in terms of income. The median family income at the time was about $50,000, meaning half the families in the country had incomes below that figure and half had incomes above that figure. We were always comfortably in the lower half. In fact, as a result of the challenges of running our own small business there were some years when our income dropped dangerously close to what in Canada was deemed the poverty line. Canada has a very generous definition of what this income level is, but I share this to show you that you don't need a high income to become financially independent. You need to be frugal and you need to get out and stay out of debt. But you have to be solely focused on this one goal. A weekly trip to the mall will not help you in this cause.

Each year as the car got older and older and we wanted a new one, we held off. Eventually whenever we took a long trip we'd rent a car. Owning and operating a car costs many thousands of dollars a year in taxes, insurance, maintenance, and repairs. We took collision insurance off the car because it was so old that if we had been involved in an accident they wouldn't have given us much anyway. But it was paid off and we elected to save a huge amount of money and keep driving that older car and put the savings towards the mortgage. Keeping the beater on the road and renting a car for longer trips that required a more dependable car saved us thousands of dollars.

For three more years we continued in our single-minded mission of paying off our mortgage. Snowsuits got used a third year when they were probably tighter than the kids would have liked, we came out at the winning end of our Automobile Club membership as we made greater use of tow trucks and services than someone with a newer vehicle, and vacations were canoe trips in provincial parks.

The other brilliant part of paying off a chunk of principal on your mortgage periodically is that as you continue with your regularly monthly or weekly payments a greater percentage is going towards the principal. Mortgages are front-end loaded, so that in the early years of the mortgage you are paying mostly interest and in later years you are paying mostly principal. So what you are doing with these regular principal payments is moving up the time continuum so that a larger percentage of each payment goes towards the principal.

Let's say you had a mortgage of $240,000 with a term of 30 years and an interest rate on the loan of 6%. The national average is that people sell their home every 7 years. If you did this you would owe about $216,000 on the principal after having paid $120,000 in mortgage payments. Of that amount only $24,000 would have gone towards the principal. You would have paid $96,000 in interest.

So **Rule #2** is to have a mortgage that lets you pay off a percentage of the principal periodically. Have this option and use it!

On the 5th anniversary of our mortgage we were able to apply one last check and pay off our mortgage. Of all the feelings in the world, there are few that rival the feeling of leaving your bank without a mortgage. We photocopied our mortgage and had a ceremonial burning in the fireplace that night. (We thought it was a good idea to hang onto the original just in case.)

It was as though a huge weight had been lifted from our shoulders or the storm clouds had left our home and the sun had come out. Suddenly anything was possible. And it's time you experienced this same feeling of freedom. It's going to require sacrifice though. It's going to mean keeping that downhill ski equipment a couple of years longer than you'd like. Heck, it means not downhill skiing at all. It means heading to the local ski swap or reuse centre, picking up a pair of used cross country skis, and finding a forest or trail near your home to ski on for free. It's cheaper and better cardiovascular exercise.

It means getting used to the fact that the vehicle you own will no longer be classed as "late model." It means that the next time your starter motor goes, rather than taking it to the dealer you're going to find a local mechanic who can put in a rebuilt starter. Or better yet it means asking your neighbor who's always out working on cars to coach you on installing a starter motor. I find most people love sharing knowledge. You can do it, and you'll save hundreds of dollars by buying a used starter from a wrecker and putting it in yourself. In the past you've traded your labor for an income that you used to pay someone else do tasks like this. Those days are over. You need to keep that income yourself and save the money by learning a new skill set.

It means saving every penny you can and putting it into the "Five-Year Plan — Pay Off the Mortgage — Move to the Dream House in the Woods" account. You need a separate account for this and you need to celebrate every time the balance goes up. Saving money in our consumer-obsessed culture is a huge accomplishment and you should be proud every time you're able to do it. But don't celebrate with a trip to the local roadhouse and end up $75 poorer by the end of the evening. It's time you learned to make pizza at home and indulge in a six-pack of beer, preferably beer you brewed yourself to save money!

As a consumer you've learned to celebrate earning an income by coming home from a shopping expedition with a bunch of shiny plastic bags filled with "stuff" to show what you've accomplished. When you don't buy "stuff" you don't end up with a little prize. You have done something much harder and far more celebration-worthy, but it comes in under the radar. So start a ritual. Every time the "Five-Year Plan" account goes up by $500, it's pizza and beer night! Every time it goes up by $1,000 it's off to the local second-hand bookstore with $10 to spend on anything you want! Every time it goes up $5,000, well, the sky's the limit! It's rent-a-new-movie night. This is a big deal because up until now you've had to wait until it hits the "Two for Tuesday Night" cheap shelf. Go ahead! Spend $5. See it as soon as it's released! You've earned it!

We've all seen those huge thermometers that hospitals and public institutions use to track fund-raising progress. It's time to make one for your fridge. Set the goal. It might be $20,000 that you want to raise this year to put towards your mortgage on its anniversary. You should transfer $1,000 of each paycheck to that account, and mark your progress each payday. Look at it every time you open the fridge. This helps you stay focused on the goal. It also gets you thinking: All right, if I can keep putting in $1,000 a month, we're still $8,000 short of the goal. Where's it going to come from? What about that guitar I don't play any more, the one Rick Nielsen of Cheap Trick signed? Wonder what I could get for it on eBay? And those Mad magazines from the 60s gathering dust in the basement? Maybe it's time to see what they're worth. And Tom down the road needs some help installing hardwood floors on the weekend. Maybe it's time I started helping him out to earn some extra cash.

If you've got a mortgage you want to pay off, you've got to be obsessive about it and a thermometer on the fridge is going to keep it in your face. If you've been having back problems and you've convinced yourself that a $5,000 hot tub is the solution, being constantly reminded about how much that purchase would set you back from your goal may finally get you down to the library to borrow some yoga tapes and get you doing some stretching exercises first.

If you're reading this and you don't yet own a home, all the better. This housing market represents a tremendous opportunity for you. The glut of homes built during the housing boom will mean that there are more places to rent, which should drive rents down. It also means that when you've saved up enough to buy a home it will finally be affordable. I am shocked at the prices of houses. Part of this is the economic environment created by Alan Greenspan and policies such as mortgage deductibility in the U.S., which encourages home ownership. One of the other factors is how large houses have become. The average American home has doubled since the 1950s. In 1950 it was 983 sq. ft., in 1970 it was 1,500 sq. ft. and by 2004 it was 2,349 sq. ft. We raised our daughters in an 800 sq. ft. bungalow that had two bedrooms and one bathroom. They shared a bedroom and in their twenties they're still the best of friends.

So if you're hoping to someday own a home, don't rush it. When Ben Bernanke took over as Alan Greenspan's replacement as Chair of the Federal Reserve he was on record as saying we had never experienced a cross-country decline in house prices. Because it had never happened, he couldn't conceptualize it. But it has come to pass with a vengeance. House prices are in free fall across the country. So the longer you can wait to buy a house the better. You should still save your money with the same zealousness as someone trying to pay down a mortgage; you're just increasing the size of your down payment, which will mean less principal on your mortgage and therefore less interest to pay in the long run. The more you put down, the smaller the principal amount, and the faster you'll be able to pay it off.

I believe house prices are very inflated and have a long way to go before they bottom out. It's important you make sure to track information from sources that don't have a vested interest in the market hitting bottom and turning around. Organizations in the real estate industry have a strong incentive to want to interpret data positively, so make sure you're using multiple sources of information to determine when you think the market hits bottom. With 600,000 to 700,000 people losing their jobs every month in the winter of 2009 it's unlikely that there will be any good news in the housing market for quite some time. Even if the economy does turn around there will be many people hesitant about entering the market until they're sure they have job security and stability.

For someone struggling with a mortgage today that is underwater, where the value of the home is lower than the mortgage, it is difficult to offer advice in an ever-changing environment of government bailouts. The reality is that if the value of your house is now significantly lower than your mortgage, you have to ask yourself some serious questions. Do you see the situation correcting itself soon? If not, for how long do you want to keep putting money into an asset with a declining value? It's always hard to walk away from something you've put a lot into. A home is even more difficult because it represents so much more than an asset. It has memories. It has sentimental value. Milestones in your life happened there. But these are not ordinary times and sometimes very difficult decisions must be made.

Walking away from a home is not something to take lightly and you should consult with a lawyer so you know all the ramifications, especially how it will affect your ability to purchase another house in the future. The implications of this vary from jurisdiction to jurisdiction, so consult with a local attorney and accountant to get all the facts. There are some companies now offering to assist you in this process for a one-time fee. The fees, which can be as low as $1,000, may sound good, but you'll probably find that an exploratory meeting with a lawyer and account will cost you less and you'll get much better advice. Laws that govern insolvency are varied and potentially changing in this challenging time, so consult an expert. Some of these companies are charging between $2,000 and $4,000 for this service and many are scams. If they tell you to write them a check and stop writing checks to your bank, be cautious. Get a professional for a decision as important as this. The www.makinghomeaffordable. gov/ website shows the various programs the Federal Government has to help you stay in your home. Make sure you research every government program that is available too. These are constantly changing so invest some real effort in finding out about what's available.

The reality is that if you have a $700,000 mortgage on a house now worth $350,000, it becomes questionable how long you should continue to throw good money after bad. I do not think the housing market will recover for a long time. During the housing boom too many homes were built, and the prolonged economic downturn will reduce the number of potential buyers for many years to come. Speculators who never actually lived in the homes they purchased fueled part of the housing boom, which further adds to the volume of unsold inventory on the market.

If you have a mortgage and you have an income your goal is simple. Pay off your mortgage, first and foremost. If you don't own a home and are looking at a market where house prices are declining, it's better to be renting that asset for now and then owning it when the market hits bottom and begins to improve.

The "Thriving During Challenging Times" workshops I have been giving for a number of years have evolved from the renewable energy independence workshops I began doing a decade ago. One of my contacts at a college where I run the workshop saw my description of it and noticed that it included a financial component. She called to question whether I was a certified financial planner. I assured her I was not. I also explained that my financial advice was pretty basic. In fact the whole financial theme of my workshop came down to one PowerPoint message which I emblazoned across the screen in the biggest font that would fit. It was simple. The key to financial independence is to:

STOP BUYING STUFF!

Seems pretty basic, but these are three very loaded words which contradict every message we receive, or at least were receiving three years ago at the height of the housing and economic boom. Americans are consumers and personal consumption accounts for 70% of gross domestic product. That's a staggering amount and it's what has been keeping economies like China's humming along, trying to keep up with the endless demand for "stuff" for Americans to buy. Acquiring stuff you'd think would make us happy, but it doesn't. Nathan Gardels, editor of the *New Perspectives Quarterly*, says, "Things are thieves of time. The more things you have the more you have to work to get those things. Frugality is the wise use of resources."[1]

Now I know what you're going to say next. "But you said I should be buying a solar thermal system to heat my hot water and solar panels for electricity." You're right. You caught me. So let me rephrase it. "Stop buying stuff—that doesn't make you more independent." Solar panels and gardening tools help you move towards independence. Designer shoes, sports memorabilia, a 1960 Ford Mustang, a riding lawn mower, a _____ (fill in the blank with about a billion useless things we all buy), don't make you independent. They just fill your life up with clutter. We now have TV shows on how to get rid of stuff. Oprah has shows where the experts come into a big home where there's no place to sit or lie down because every square foot is filled to the roof with "stuff"! After dozens of dumpsters are hauled away and they hold a garage sale that occupies an entire warehouse, the two people that live there can return. Does that not strike you as pretty bizarre?

At the same time that American houses are getting bigger and bigger we have this whole industry of storage units where people take even more stuff to store off-site. How much "stuff" can one family own? This is part of the reason we're in the mess we're in. It used to be that people saved some money. The bank took that money and lent it out to other people in the form of mortgages so they could own a home. Now banks often don't have enough savings on hand to do that and have to look elsewhere for money to lend. And some of the money that comes in to support our North American lifestyle comes from China, where people still save money.

So you need to return to those days when people saved money. You've now got at least two new bank accounts open for specific goals to save towards. With separate passbooks at separate financial institutions that you can ride your bike to conveniently. You have the "Solar Power" account where you're putting all your energy savings to buy solar panels for your roof. You've also now got a "Five-Year Plan/Pay Off the Mortgage" account where you're putting every other penny you can save.

Credit Card Debt

One of the things that are going to keep you from achieving those savings goals are those magical pieces of plastic in your wallet. Credit cards are incredibly convenient and a great invention. But they're like many other inventions. Morphine is an incredible painkiller, but if you use it too much and for too long you get really addicted to it. In the early stages of its use it serves an amazing function: preventing extreme pain. Eventually, though, if you don't wean yourself from it it becomes as big a problem as what you needed to take it for in the beginning and you become addicted.

Credit card debt is an absolute no-no and you need to eliminate any credit card debt you have. It's much worse than mortgage debt because with a house once you eliminate that debt you have an asset that has great value. You can live in a house. It keeps you warm and comfortable and for many decades it maintains its value. The stuff you purchase with your credit card generally will not maintain its value and you've purchased a depreciating asset. If you've been carrying a balance on that credit card to pay for these things that lose value you're burning the candle at both ends.

Paying credit card interest is to your household financial health what smoking is to your personal health. It's bad and there is no upside. Ultimately it's going to severely weaken your ability to thrive financially. So you've got to pay it off. I'm hoping you can do this fairly easily and quickly. I'm hoping you have a couple of credit cards and a couple of thousand dollars of debt that you can focus on and get rid of even before you focus on your mortgage. My concern is the trend that I've seen recently on several TV shows that focus on people with severe debt problems. Sometimes they have 10 or 15 credit cards and they have tens of thousands of dollars' worth of debt. If that's the case you have to take drastic action. You have to go on a crash diet and give up everything but bread and potatoes until you pay them off. Sell the car, sell the motorcycle, sell anything and everything and don't buy anything until all your credit card debt is paid off.

If you were hoping for some magic bullet to make this disappear, I'm afraid I don't have one. A huge balance owing on multiple credit cards is going to take some real time and commitment to get rid of. And you have to because the interest you are paying on credit card balances is very high. It is much higher than your mortgage interest and higher than a line of credit which is secured against assets and therefore usually offers better interest rates.

Annual interest rates range from 8.99%, which you would pay if you had an excellent credit rating, to up to 30% for someone without a good credit rating.... OUCH! Let's take the example of purchasing an item for $1,000 on your credit card and only paying a small or minimum amount each month. If you had a good credit rating and were able to get a credit card with a low annual interest rate of 10% and you paid $20 per month, you would end up paying $300 in interest and it would take you 5 1/2 years to pay it off. If, however, you did not have a good credit rating and your credit card had an annual interest rate of 20%, that same $1,000 purchase at $20 per month would take you 9 years to pay off and you would end up paying $1,171 in interest! If you were paying a rate as high as 25% and you only paid $20 per month you would never pay the balance off. After 10 years you would have paid several thousand dollars in interest and still owe the original $1,000.

So if that $1,000 item costs you $2,171 ($1,000 + $1,171 interest) does this sound like a wise thing to do? Wouldn't it be easier to save the $1,000 and purchase the item for cash? Credit cards are a trap, just like the quicksand you used to see people fall into in old movies. Don't get caught in the quicksand or credit-card trap.

Saturday Night Live recently did a brilliant sketch about consumers struggling with consumer debt. The husband and wife are sitting at the table trying to figure out how to pay off their credit card bills when the author walks into the room with his new book *Don't Buy Stuff You Cannot Afford*. They are perplexed by the title. Amy Poehler, the wife, reads from the book: "If you don't have any money you shouldn't buy anything. Hmmm. Sounds interesting." "Sounds confusing," says Steve Martin, the husband. The book recommends buying things with money you actually save. The husband asks, "Let's say I don't have enough to buy something, should I buy it anyway?" "No" answers the expert and this goes on for several minutes with the couple unable to grasp the concept of living within your means. Sometimes it takes comedy writers to make crystal clear something like the lack of wisdom in the way we've all been living for years.

Under their "Consumer Tools" the National Foundation for Credit Counseling (www.nfcc.org) has some excellent calculators that allow you to plug in your account balances and interest rate to get an idea of how much interest you'll end up paying. If you have debt right now go to this website and get a handle on the reality of what it's costing you.

You can see that paying $300 interest on a $1,000 purchase is a lot of money. Paying $1,171 in interest on that same purchase is insane. It's bad money management. You simply have to stop using credit cards in this way. If you need to buy a new solar panel to add to your system, you need to save the money to do it and pay cash for it. If you've researched solar panels and have a pretty good idea of what they're worth and you happen to be at a renewable energy fair and a dealer is putting away her booth at the end of the show and offers you the panel at an exceptional price to save having to truck it back to the shop and you don't have cash on you, then it's all right to use your credit card. You have less than a month to come up with the money to pay off that amount on your credit card bill when you get it. This is the way you have to think about credit cards if you have them. They are a temporary tool, not part of a sound financial strategy.

If right now you have multiple credit cards with balances owing, pick the card with the highest interest rate and pay it off first. Then take out those scissors and cut it up. Celebrate it. Have a pizza party. A homemade pizza party. Then take the card with the next highest interest rate and pay it off. Many financial advisors today say that once you pay it off you should put it away and just not use it. This is because your credit rating will be better if you have lots of credit cards. They suggest you just use it once in awhile to keep it active but pay it off. I completely disagree. Credit is a drug and if you've been paying interest on a credit card you're a junkie. So stay away from the dealer and cut the card up. If you were a heroin addict do you think it would be a good idea to keep some in the fridge just to prove you weren't addicted and therefore would never fall off the bandwagon?

If you have multiple credit cards with balances some credit counseling services suggest you try and get a consolidation loan to pay off the credit card companies and roll all that debt into one big pot. Then you can begin paying off just that loan. The interest rate will be sane and you'll get out of that spiraling trap of compounding interest on interest. If you do have multiple cards with balances it may be worth asking a credit counselor to take a look at where you are and provide some feedback. The National Foundation for Credit Counseling is a non-for-profit resource for this and can be reached at 800-388-2227 or nfcc.org.

The biggest factor in success with using an outside support person is that often they'll be able to step back and take a big-picture look at your situation to make sense of it. When you're directly immersed in the mess sometimes it's difficult to be objective about the solution. It's like having your mom or dad tell you to do something when you're a kid. Sure you're an adult now, but if someone behind a desk in an office is what you need to get your credit card debt in line, then by all means, get one on your side.

Many years ago I went to an investment counselor to invest in the stock market. He went through an inventory of our assets (none, we rented an apartment and had a car loan) and our incomes, which were moderate at that time. Then he asked how much we had in savings. "Savings?" I asked. He said, "You make that much money and have no savings? Get out of my office and don't come back until you do!" I think I was about 25 at the time but felt as though I were 7 and my father had just yelled at me for setting fire to my plastic soldiers in the sandbox. I was mortified. From that day on Michelle and I started saving money, and we've been savers ever since. That's the advantage of getting advice from an outside person. Often you'll do a better job of getting your house in order for them than you will for yourself. I'm sure there are many books written about what in our makeup causes humans to behave like this. I'm guessing it goes back to pleasing your parents, but that is another book. But who cares. Once I started saving it was like: "Ken (my financial advisor) will be so proud of me!" And sure enough six months later we went back and he was impressed and decided to take us on as clients.

If you have manageable credit card debt, say $2,000 to $5,000 on a couple of cards, you should just put your nose to the grindstone and pay them off quickly. With the interest rate you're paying on a credit card it has to go first, and it has to go soon. You can't start making any plans about independence if your income is being sucked into the vortex of credit card interest.

I created the analogy of credit card debt being like smoking and nicotine to your financial health, but I think nicotine may be a little tame. Credit cards themselves are like any of those addictive drugs like cocaine that we hear so much about. Why? Because they create an illusion. Cocaine creates an illusion of well-being. Credit cards create an illusion of financial well-being that isn't reality. They allow us to walk into a mall and think that, regardless of our income, anything is possible (as long as it's less than our credit limit). It creates the illusion of infinite abundance when for just about everyone money is finite. It has limits based on our income and assets. Credit cards let you break free of the bounds of gravity imposed by your income and soar to unimaginable heights. But just like Icarus and his wax wings, when we soar too high we end up being brought back down to earth, often with a thud.

As with cocaine, it's not easy to break the addiction. If once you finally get that credit card paid off you don't think you can stay clean and not ever let that balance get out of control again, you should stop using credit cards. You should cut them all up. Again here I'm going against the conventional wisdom of so many financial advisors who suggest you should keep the cards and just make sure you pay off the balance each month. This helps with your credit rating. This sounds great; it's important to have a good credit rating I guess. It's good if you plan on continuing to use credit for the rest of your life, but I'm not taking you down that path. Buying stuff to make yourself happy doesn't work. It's the wrong path. I'm suggesting that your goal is financial independence where you don't need a mortgage and a line of credit and credit cards to get by. You need to change your perception of reality and get off the credit bandwagon, and getting rid of your credit cards is a good place to start. Pay 'em off and cut 'em up!

Until recently one of the advantages of credit cards was the convenience of being able to substitute them for cash and not having to carry so much cash with you. But debit cards have changed that. Debit cards are much more fiscally responsible because they are like cash. The retailer you're completing the purchase with simply removes the money from your account electronically. If the money is not there, then you're going to get turned down. A debit card has the convenience of a credit card but with the reality check built in. If you don't have the money in your account, don't buy it. So scrapping your credit cards and going to a debit card makes sense, although if you know you have the resources to pay for the item, it's nice to have a credit card to fall back on. Having one credit card that you only use for emergencies and always, always ALWAYS pay off completely each month is also not such a bad idea.

I have one credit card. I use it because it's convenient and because I get reward points from a hardware retailer I like who also sells gas. As much as I'm trying to reduce my carbon footprint, I still drive and I still buy gas. This retailer is rewarding me so I buy it there. I don't buy more than I need in order to earn points. I just earn points on what I'd be buying anyway. And since I get to spend my points in a hardware store, I can always find stuff that helps me in my quest for independence, whether it's tools to fix things or gardening supplies. I know what you're saying. I'm part of the consumer culture! I'm getting rewarded for being a consumer! I'm a hypocrite! OK, you got me. But I still live in a capitalist society and it sells a lot of amazing products that make my life easier and allow me to be independent. Gas is gas. It's just a commodity so I'm putting gas in my tank and this retailer is going to give me some points I can use towards a new cultivator for the garden. I'll go there.

Line of Credit

A line of credit is a vehicle that allows you to borrow money against a fixed asset like a home. As the housing market was on its way ever upward early in the millennium many people got very familiar with lines of credit. With the value of homes increasing, a bank or financial institution was often willing to loan you money to purchase things because the loan was secured against the house. The mortgage holder would be first in line to get paid if you defaulted, but the line of credit was often issued by the same institution, so it was just another way for the bank to make more money. You paid them interest so that you could have something immediately rather than waiting until you could afford it. This in itself is not a bad thing.

Our family has a holiday tradition of watching *It's a Wonderful Life* on Christmas Eve. This is an outstanding movie on so many levels. Jimmy Stewart does an exceptional job; I'm always amazed at any actor's ability to memorize lines and then deliver them with such conviction. In one scene Jimmy Stewart's character George Bailey is ranting to the evil banker Mr. Potter about how his father's Savings and Loan Company had loaned money to people the bank had turned down for mortgages. Potter suggests they shouldn't own a home until they can pay cash. George Bailey disagrees and suggests it's really not such a bad idea for a family to have the money to purchase a home when they need it while their families are young rather than wait for 25 years while they save and buy it when the kids have left home. It gives them dignity and pride in where they live and a sound financial asset at the end of the process. "Well, is it too much to have them work and pay and live and die in a couple of decent rooms and a bath? Anyway, my father didn't think so."

I agree with George Bailey. Using a mortgage to buy a home for a family is a wonderful part of the financial system. It's an admirable thing to allow a young family to take pride in home ownership and end up with a sound financial asset. Where we might be creating a problem though is with lines of credit. As housing prices kept going up, lines of credit allowed forever-increasing expectations on the part of homeowners in terms of what they could have. Want a boat? A nice big boat? No problem, use your line of credit. Want a high-end SUV and can't afford it? Forget about it. Use the line of credit. Always wanted a designer kitchen just like the ones on those TV shows? It's as easy as writing a line of credit check. People were able to use their homes as bottomless ATMs and just keep withdrawing cash that was appearing out of nowhere as their house magically went up in value. It was "easy come" and now we're watching the devastating impact of "easy go."

The problem with lines of credit is that financial institutions took a pretty sound financial concept, a mortgage, and pushed it beyond its logical limits. People really didn't need yachts and high-end cars and designer kitchens. They could buy a canoe with cash until they'd saved for the yacht. A yacht isn't a necessity. They could drive the Chevy until they could afford the Cadillac. They could still eat and prepare meals in that dated 1960s kitchen until they'd saved up for the new one. Like so many things in our time we had to push the limits. We had to turn credit into an extreme sport and see how far we could push the envelope. When the housing market collapsed, that envelope disappeared pretty quickly for most of us. And reality set in. Guess what? We really couldn't afford the yacht after all on our income. In fact, by the time we got through paying all our other expenses we couldn't afford the gas for it. So I guess the canoe was a better idea after all.

Somehow our instant-gratification culture forgot that sometimes good things are worth waiting for. It's not necessarily realistic for a young couple to own a massive home immaculately furnished and with two huge

SUVs in the driveway. Remember the olden days? The 50s and 60s and even the 70s where you rented until you could afford a half-decent down payment? And when you did buy that first tiny fixer upper, your voice echoed around inside because there was so little furniture to absorb the sound? Remember the crappy linoleum floor and tacky countertop and stained sink? People who are now in their 50s grew up in those houses and really didn't mind it. There weren't as many high-tech gifts under the trees but a new tricycle and a plastic gun could bring such joy to your life. Remember how that tree with the rope swing could amuse you for hours? How is that possible? There was no monstrous wood and plastic climbing structure with a built-in play house and climbing wall, and yet somehow most of us were pretty happy as kids.

So the line of credit has to go as well. Or at least using it for frivolous purchases has to stop. A solar domestic hot water system purchased on your line of credit when you know will be paid off in three to five years is probably not such a bad idea. You get hot water from the sun and as long as you pay down that line of credit you end up with an asset that's going to reduce your carbon footprint and save you buying natural gas for as long as it works. That's a good use of a line of credit. A granite countertop will not help you achieve financial independence. A solar hot water system will. Some government programs require that you do energy-efficiency and renewable-energy upgrades within a fixed time period after an energy audit. If you don't have the cash to complete everything before the time expires but there is a significant rebate if you get the system installed, use the line of credit. Get the solar domestic hot water panel installed and as soon as that government rebate check arrives in the mail put it towards the line of credit. This is an example of a line of credit being a useful tool in your quest for independence.

So if you have a line of credit with nothing on it that's a good thing. Just leave it there. Don't use it unless you can really cost-justify the purchase of an asset. If you have a balance on it, pay it off. Like credit card debt the interest rate will be higher than your mortgage, so pay it down before you get cracking on your mortgage. The order should be to pay off your credit card debt first, your line of credit next, and then your mortgage. Some people may find they have a great rate on their line of credit and may want to use it to pay off all those credit cards. Just remember that you then have to double your efforts to pay off that line of credit. It's secured against your house. If you can't pay off that line of credit, remember that the asset that secures it is the one you call home.

So get serious and pay it off.

Forced Savings

Many of use are familiar with retirement plans where money is put into our retirement fund directly from our paycheck, even before we see it. This is forced savings. While you could get your paycheck and do this yourself, for many a plan like this ensures that the money gets where you want it to go. Well now that you've made the commitment to be more financially independent you have to take it to the next level. This means creating a forced savings account. These are after-tax dollars, which means the money you get from your paycheck. It ends up in your bank account and there's a huge temptation to do something else with it. Buy a big-screen TV, take a trip to Vegas, get that motorcycle you've been thinking about. Forget about it. They're not going to happen. You need to have discipline and immediately put that money into your forced savings account.

You can prioritize and decide what the account is for. It can be the "Solar Power" account, the "Five-Year Move to the Country Plan" account, or your "Emergency Fund" account. Whatever it is, it has to become habit that you put money into this account first. It's the "Pay Yourself First" budget plan. Before you pay the satellite TV bill, pay yourself. If by the time that satellite TV bill is due you don't have the money, you're going to have make a choice. Are you going to put it on a credit card? We now know you're not. Are you going to not pay it? That's not the best option since they won't be happy with you. Or are you going to call them, cancel the service, pay off the remaining balance, and hook up that antenna that your grandparents have stored in their backyard? Oh, you won't have the number of channels or the picture quality, but you will get your TV reception for free. This is what we're talking about. You've got to make hard choices if you're serious about becoming financially independent. There are no easy tricks.

You've got to spend money in the most effective way possible. As William Kemp notes in *$mart Power*, you're going to spend close to $250,000 or a quarter of a million dollars in your lifetime on energy. This is electricity to power your home, natural gas or oil for heat, and gas for your vehicles. That's a lot of money. If you follow his payback chart on page 108 you can see how to reduce those expenditures through energy efficiency. You can start taking some of your after-tax dollars and purchasing energy-efficient appliances and putting the energy savings into one of these forced savings accounts. This is a sound use of your money and a viable reason to consume. In this case, the acquisition of some product is moving you towards your goal of increased independence.

Emergency Funds

The first forced saving fund you'll be setting up is your "Emergency Fund," which is designed to provide your family with a cushion against the loss of an income. The amount varies and it's actually an irrelevant number because there are so many variables. Some financial advisors use eight months as the number. Some say three months. How about a year? You should calculate a monthly budget and once you know how much money you need to cover your expenses for a month, from the mortgage payment to food and utilities and insurance, multiply that by eight months and that's how much of an emergency fund you should have.

Obviously the lower your monthly expenses the smaller your fund can be, which is another great reason to reduce your levels of consumption. The smaller your home and the more affordable your insurance and expenses the easier time you're going to have in the event that someone in the house loses their job. So by all means have an eight-month emergency fund. Have a year-long emergency fund. Ultimately, though, what you're working towards is the day when the loss of a job won't be as big a deal because you'll have your mortgage paid off, you'll not have debt, and your living expenses will be minimal. Installing a solar hot water system will mean less natural gas or electricity for hot water. By installing solar electric panels you won't require so much grid power. The goal of paying down debt has the added bonus of increasing your financial independence and lessening the impact of job loss.

So pick a number—6 months, 8 months, 12 months, whatever works for you—and set that money aside in a government-insured savings account. This shouldn't be considered a financial asset, just a safe and secure, easily accessible savings account. If you can earn some interest on it great. But growing this asset is not the goal. Keeping it safe for when you need it is paramount. As you're developing these forced savings accounts—for emergencies, renewable energy equipment, your yearly mortgage principal payment, and your five-year plan—give some thought to having them at different financial institutions. This spreads the risk of a bank failure. It also increases the pleasure you'll start taking in saving money. Each payday involves a trip to two or three different institutions on your bike to make a deposit into a savings account. When you leave you can look at the passbook and see how you're doing. People like to keep score and this is an excellent way to do it. If you just have one savings account you use for multiple purposes it won't have the same impact and it becomes difficult to keep track of how much of the total is for each goal. Having separate bank accounts in different banks is just way more fun.

Monthly Budget

I've already mentioned the necessity of a monthly budget to help you calculate the amount you'll need for your income loss emergency fund, but it's worth repeating just how important it is to actually have a budget. Most people have computers, so there's no excuse. You don't even need a computer for that matter, although there are lots of inexpensive home budget programs if you don't want to just make your own on a spreadsheet. Sometimes doing it on a computer makes it more inviting and therefore more likely to get done.

Budgets serve two purposes. The first is knowing what you need to live on. The second is to expose just where you're spending your money. I think most people are surprised when they actually see where their money goes. Sometimes you think you have an idea of how much you spend on things like groceries, but then when you see the actual amount you spend you're flabbergasted. That's one of the beauties of a budget. Having accurate information about your finances is an excellent tool for everything else we've discussed in the book. You won't know whether you can afford a new wind turbine unless you know where you're at, and using the seat-of-your-pants method where you guess whether you have the money or not doesn't work anymore. Credit is tighter than it used to be and you've seen the necessity of living within your means, so putting it on the credit card doesn't work. That purchase will have to wait.

So set up a budget and stick to it. First set up all your income sources, then expenses that are withdrawn from your account monthly, like insurance and car payments.

Keep all your receipts and start inputting them into the budget. It's crucial that you track everything from your morning coffee to magazines you pick up for the subway ride home. Two $5 coffees at work is $50 a week or $2,500 a year. It doesn't seem like a lot each time, but two years of making your own will almost buy you a solar domestic hot water heater. Then half your hot water is free every year and instead of sending that money to the gas company you can use it to pay off your mortgage. It's all about having good information. You can't make good decisions without information about money in and money out.

After a few months sit and down and analyze your budget. What's going well? Where could you improve? Whoops, we spent $400 at restaurants last month. We need to cut that in half. Someone spent way too much money on clothes. Time to find some local second-hand shops and reduce that. We spent THAT much on food? All right, why did we spend so much? Too much of it was processed and ready to eat. So we need to start making more meals from scratch. There will be an infinite number of ways to reduce expenses once you actually know how much you're spending.

Investing

Oh great, now we get to talk about investing! This is the best part. This is the part most financial books talk about. In fact, there are tons of books on just this topic! So I think I'll leave those books alone and give you an entirely fresh perspective on investing. I think right now what you need to do is forget about investing, forget about growing your assets and looking for a reasonable return on investment, and concentrate on capital preservation. Wow, that sounds complicated! Well not really. I'm just suggesting that rather than trying to make money you focus on not losing it, and in this market that's a real challenge. In fact with the way the markets have been going, it's been almost impossible to just stand still, let alone move forward.

There comes a time in every market when it's time to sit on the sidelines and let things settle down, and now is one of those times. If you ask when I think you should get back in I'm going to suggest not for a long time. In fact, I'm not sure if you're ever going to want to get back in. I think the market is going to continue to go down for a long time and it's going to take a long time for it to come back. "But how am I going to grow my money so I can retire comfortably?" Well that's the $64,000 question and to be quite honest I think we're all going to have to rethink the concept of "retirement."

Right now we all envision saving our money in our retirement plan, watching it grow year after year, and then when we can afford it investing that money in a financial instrument that will keep paying us a reasonable amount each year, enough to live comfortably on. For a lot of people close to retirement, that vision has just changed because the money they invested lost so much of its value as the stock market underwent a precipitous drop. They simply don't have the amount they need to retire.

Or at least they don't have the amount they need to retire in the comfort they had expected. A lot of us have high expectations for that retirement. Wintering in warm places. Days spent on a golf course. Regular cruises, lots of trips exploring exotic places around the world. Meals out, new cars, cottages where the kids and grandkids can visit us. You know, just like we've all seen in the financial industry ads. They look great! Count me in! I could never take enough trips to exotic places!

The reality for many people up until now is that their retirement never lived up to those images anyway. With the financial crisis the reality of retirement has now changed for everyone else as well. It's simply not going to happen for many of us, and for many others retirement is going to happen in a much more diminished way. I hope this doesn't sound too cold-hearted, but I don't see this as a bad thing. I've always found the concept of retirement a little depressing. I guess this comes from some of the things I see seniors doing, like mall walking and sitting on porches looking bored. They just don't seem to have enough stuff to do and can get pretty bitter about it. Perhaps it's because I know of so many people, men in particular, who defined themselves through their work, then retired, and died shortly thereafter. I don't know whether there's any empirical data to back this up, but my first-hand experience tells me that with some people it's the case.

They say youth is wasted on the young. In the same sense retirement is wasted on the elderly. When I was younger there were so many things I wanted to do, all of which involved lots of time, which I didn't have. I had a young family. I was starting my career. I was putting in long hours. I was working hard to pay off my mortgage. Once you get to retirement age you just don't have the stamina to do so many of the things you dreamed you'd do when you had the time. It's the beauty of the movie *The Curious Case of Benjamin Button* where the lead character is born old and grows younger as time goes on. It makes complete sense. Now we just have to come to grips with the fact that we are still going to be accomplishing valuable tasks as we age and we're not going to have the opportunity to get bored. In so many cultures elders are a revered and integral part of the family. In ours they're often a burden and shipped off to seniors homes and supervised care.

As we all adjust to the new economy, one with fewer jobs and a re-duced level of economic activity, money will be harder to come by and sharing accommodation and sharing family responsibilities will become more and more important. Working parents who have had their parents close enough to provide child care have been very fortunate up until now. The necessity for the older members of the family to help with the younger is going to become more important than ever. This is the way families used to operate, and in the new economic order families will often need to return to this method of operation. This is the beauty of the concept of humans organizing themselves into these units we call "families." Sure, lots are dysfunctional, but the concept is still good. Provide a support network, nurture the young, support the elders: it's a pretty sound idea.

Families being closer together is actually a really, really great thing. Think of how wonderful it will be to value our seniors once again. To have their wisdom and experience close at hand. Think of how much grandchildren will learn spending more than just Christmas and birthdays with their grandparents when everyone's stressed out and behaving badly. And think how much more valued and loved seniors will feel when their contribution to the family is recognized and appreciated. All over the world different generations live under the same roof, and during much of the history of North America economics dictated this lifestyle. We are simply returning to a system of living that we have only moved away from in the last few generations. I am always amazed at how many people I meet whose kids live in Seattle, San Francisco, Boston, or Paris. Cheap and abundant energy has allowed this wholesale dispersion of families. I believe that with the ongoing economic malaise and the end of cheap oil there will be a huge incentive for families to get closer together geographically.

I'm not suggesting that all families will have to move in together. Some people will still be able to retire as before. Some may have to continue working longer than they would have preferred. Some will have to take on part-time jobs to make ends meet. But none of these things is necessarily a bad thing unless you choose to perceive it that way. Having a sense of purpose makes people happy. Contact with others and interacting with younger people helps keep older people mentally engaged. Tending a garden that reduces the family's expenses is something that is going to become really important and it's going to make senior members of the family feel very happy to be making such a huge contribution. Exercise is what keeps people young. Load-bearing exercise is what doctors recommend to fight osteoporosis. Being outside nurturing the garden is a far better way to spend your golden years than watching reruns of *The Golden Girls* all day.

Frankly I can't think of a worse fate than to get up everyday with nothing to do. Nothing to contribute. Just filling my time, taking up space on the planet. Seniors are an incredible asset to society, and we need to make good use of them. As I approach my 50th birthday I can really start to understand how elders have been so revered in many cultures. As you get older you just get things you never got as a younger person. You have years of experience, you've read hundreds of books, you've acquired knowledge from so many sources that you are just better able to see things as they really are. Utilizing all this wisdom and having it continue to contribute to society in the challenging times we're in for is even more important.

This idea that the concept of retirement as a time of leisure and luxury is approaching its end is not going to sit well with a lot of people. We've really only had one generation where it was the norm. The generation born after the Great Depression was basically the first and only generation that will get to experience retirement as we know it today. Many groups in society worked hard to build up pensions. Autoworkers fought for it, government workers demanded it, and in the 50s and 60s and 70s companies that wanted to keep good employees offered it. But we saw globalization move jobs offshore and the security of full-time and lifelong employment being replaced by contract work and jobs in developing economies. During this economic meltdown all pensions have suffered a real setback. Much of the money that had been put aside to pay for pensions was invested in equities and it has suffered huge losses. People who had personally controlled retirement funds were lulled into a sense that stocks would always go up, and since they had never suffered a dramatic drop in stock values they couldn't conceive of anything else. Now reality has set in. What goes up can come down, and sometimes comes down hard.

So what is my advice for investing for your retirement? My advice is to start living in the moment and enjoying yourself today because projecting what the world and the economy are going to look like in 10 or 20 or 30 years is a fool's game. It won't look the way it used to and it won't include limitless growth and opportunity. I suggest you start creating a lifestyle that you can continue to enjoy for a very long time. Living frugally. Reducing your expenses. Growing your own food. Powering your own home. Living within your means. Reducing your demand for "stuff" and therefore your footprint on the planet. We all need to experience "simple abundance" and start to enjoy simple pleasures like harvesting carrots from the soil we have worked hard to condition.

I think you need to first get to a stage where you are debt-free. That has to be your first goal. This is going to take some time if you've been living the North American Dream of spending without limits. Once you get to that stage you need to create a home that is as self-sufficient as possible, from how you heat and cool it, to where you get your electricity, to how you heat your water and where you get your food. A return to the spirit of independence and self-reliance that built this country and inspired your ancestors to come to this place. Once you become independent you can start saving money again. That money should be kept in cash or precious metals and I'll discuss this in the next chapter. Some cash you should leave in a bank that you have confidence in. It has to be backed by the Federal Deposit Insurance Company (FDIC) in the U.S. or the Canada Deposit Insurance Corporation (CDIC) in Canada. You need to be critical of those institutions and consider whether or not they can make good on their obligations, in other words whether they can pay you back if the bank fails. This insurance is backed by the Federal Government and you have to decide if you think that's good enough for you. The government has been a very stable and secure institution for a long time, but the economic crisis is putting enormous pressure on governments, which are throwing money at this problem, money that they have to borrow or create.

It's hard for us to conceptualize, but governments do go bankrupt. Governments do default on their loans. The dollars they issue are only as good as the confidence people have in them, and if people lose their confidence then those pieces of paper can become worthless. That is why my advice so far has been for you to have hard assets, things that actually do something. Solar panels make electricity or hot water. They accomplish something, they improve the quality of your life and make you more independent. They are a sound investment. Purchasing shares in a company is a soft asset. You may get a monthly piece of paper showing your ownership in that company, but it is a piece of paper that represents electronic pixels on a computer somewhere. If your statement shows that you own shares in Bear Sterns, the value of your investment went from $80/share in April 2008 to $40 a share in a matter of months, then dropped from $40/share when the market closed on Friday, March 14, 2008 to open at $2/share on the Monday morning after the forced sale to J P Morgan Chase. The same can be said for Lehman Brothers, which went from its 52-week high of $67 to bankruptcy in a matter of months. That is a paper asset and it lost its value as easily as if the paper had blown away in the wind.

Your solar panels, on the other hand, assuming you have them fastened down well and they don't blow away in a hurricane, will keep producing electricity and hot water as long as the sun comes out. The monetary value of that hot water and electricity may change, but the value of what you can accomplish with what they produce, like taking a shower or using a washing machine to clean your clothes, never diminishes. The way electricity replaces human labor is an absolute wonder to experience. We've come to assume that electricity has always been there and always will be, but we cannot take it for granted. It's just too valuable to be without.

The turbulence the world economy and world financial system is experiencing is going to continue for a long time. Harry Dent in *The Great Depression Ahead: How to Prosper in the Crash Following the Greatest Boom in History* suggests it could last until 2017. Others have suggested we are into a long L-shaped depression which is going to continue for decades. If at some point you feel the worst is over and the markets are recovering then choosing to invest in stocks again is your right. You may also want to be more conservative and invest in bonds, particularly government bonds. Treasury Bills have been considered one of the safest investments you can make. You give up the potential of a great return, but you do get security.

I cannot tell you how long the depression will last, how low the stock market will go, whether or not governments will default on bonds, and if and when things will turn around. I can tell you I believe that with the number of challenges facing the world today the shiny happy days of never-ending growth in the stock market are over. The easy money was made and now much of it has been lost just as easily. I am not advocating the easy way out. I'm not suggesting there is a simple solution to the mess or a foolproof way for you to make your money grow. I don't think there is. I think you need to change your relationship with money and reduce your expectations in terms of your financial future. This won't get me elected to office, but I think it will help me deal with a much more subdued future. One where people travel less, buy less, eat food grown more locally and support workers in their community rather than 12,000 miles away. This is a good trend for the planet and a good trend for communities. It will make them much more livable.

You may have read this chapter and thought about other things you could be doing financially. You've probably even thought of specific new products on the market that you could use to speed the process of growing your independence and wondered why I haven't mentioned them. Truth is I don't know about them. I don't because I don't need them. I don't have a mortgage, I don't have any credit debt, I have money in my savings account, I have a few bars of silver, so I'm out of the loop when it comes to the wild and crazy world of modern financial products. While you should check them and see if they can help, your financial strategy shouldn't be complicated and need long-winded descriptions. Stop buying stuff and save your money! That's all you need. All the other stuff is fluff.

All the fancy reverse mortgages and financial instruments just distract you from the basic goal of this book. To get independent you need to save more money and spend less. You need to pay off debt and start saving. It's not rocket science and the more complicated the product someone is recommending the more it may be distracting you from the basic goal. It's really simple. Stop buying stuff. Pay down debt. Save money. Period. End of story.

20   Mediums of Exchange

Once people were able to create money at virtually no expense, no one ever resisted doing it to excess. No paper currency has ever held its value for very long. Most are ruined within a few years. Some take longer.

Some paper currencies are destroyed almost absentmindedly. Others are ruined intentionally. But all go away eventually. By contrast, every gold coin that was ever struck is still valuable today, most have more real value than when they first came out of the mint.

William Bonner, Empire of Debt[1]

I spent a lot of time in Chapter 3 explaining the history of money and the gold standard. I did this because I think it's so important, because there was a time when a paper dollar was backed by a valuable asset, a precious metal like gold or silver. Today that's not the case. A dollar bill is a "fiat currency" that has value because a government says it does. It only has value as long as everyone continues to believe it has and continues to have confidence in the government that created it. The U.S. government is now spending money in unprecedented amounts to solve the financial crisis. Its current debt, the amount it has spent over time that it has not paid back, is over $11 trillion dollars. This is how much money the U.S. government owes. It's like a credit card debt. President Obama's first budget of $3.6 trillion will add another $1.75 trillion to the total this year and it's expected that this will continue while he tries to get the economy working again.

A trillion is a huge number. It's an incomprehensible number. This is what it looks like: $1,000,000,000,000. A million is a big number.

A trillion is a million million. It's a thousand billion. So $11 trillion is $11,000,000,000,000. Wow! That's big.

How can it ever be paid back? Well, to be fair I'll give the usual rationalization. Economists say that as the economy grows the government will be able to pay it off. Over time it's hoped that it will be whittled away at and be a smaller percentage of the gross domestic product or GDP.

The problem is that on May 28, 2008 Richard W. Fisher, President and CEO of the Dallas Federal Reserve Bank, estimated the obligations of the U.S. to be actually **$99.2 trillion**. And if we add the new debt, courtesy of Wall Street bankers, the obligations of the U.S. taxpayer rise to an impossible-to-repay sum of **$105,200,000,000,000.00 ($105 trillion).** Fisher is including obligations the Federal Government has for Social Security and Medicare.[2] These are called "unfunded liabilities" because they are money the government is on the hook for but hasn't put into a savings account yet. So as all the baby boomers start retiring and demanding their due from the government, it's going to have to come up with the money somehow. And George W. Bush obviously decided that older people are more likely to vote, so created the Prescription Drug Plan which added another $18 trillion to this unfunded liability. $10 trillion here, $18 trillion there, pretty soon we're talking real money. And pretty soon Atlas is going to shrug.

In the 1970s the debt was about 30% of GDP. In 2009 GDP in the U.S. is projected to be $14.291 trillion with the debt getting pretty close to that amount. U.S. GDP in the final three months of 2008 declined at an annual rate of 6.2%, which means that while the U.S. economy is contracting or getting smaller, the U.S. government is spending money and increasing its debt at an accelerated pace. How long can this continue? No one knows. I guess it can continue as long as people continue to have confidence that the U.S. government will be able to pay the debt back. If it can't, then the value of those dollar bills in your wallet could change radically and quickly. They could have a lot less value. Eventually they could become worthless. Sure that's a worst-case scenario, but sometimes it's a good idea to hope for the best and plan for the worst. The people who say that the U.S. dollar could never lose its value dramatically are often the same ones who said we would never have another depression. They say we learned our lesson the last time and would never let it happen again. And yet here we are with the worst worldwide economic crisis since the Great Depression and things continue to deteriorate by day.

One result of an economic crisis like this is that governments try and revive the economy by increasing the money supply, simply cranking up the printing press and pumping more money into the economy. While this sounds logical and easy it's dangerous because it is highly inflationary. When everyone suddenly has more dollars to spend and there hasn't been a proportionate increase in the goods and services people purchase, the people selling things start raising their prices. If more people want to buy things and the demand goes up without the supply also increasing, prices go up; this is what we call inflation. More dollars are chasing fewer goods, so everything costs more. Having more dollars in your wallet isn't necessarily a huge advantage because each dollar buys proportionately less stuff. Increasing the money supply to deal with economic downturns has been a common strategy. Many governments over the last few decades have fought the temptation to do this overtly, but there is a strong possibility it could happen again.

With the negative connotation that inflating the money supply can have, the government is now calling this "quantitative easing," but it's the same old thing—inflating the money supply or increasing liquidity. This really is one of the government's last hopes because it can no longer use interest rates to try and reignite the economy. The interest rate is essentially zero, so it has no room to move. This is exactly the problem Japan has experienced for the last two decades.

This inflation is actually eating away at your dollars, making them worth proportionately less. Some economists like John Williams (www.shadowstats.com) suggest that governments are actually reducing the value of your dollars by not honestly reporting the money supply. In 2006 the U.S. Federal Government stopped reporting the broadest measurement of the U.S. money supply, the "M3." Texas Member of Congress Dr. Ron Paul says that, "M3 is the best description of how quickly the Fed is creating new money and credit. Common sense tells us that a government central bank creating new money out of thin air depreciates the value of each dollar in circulation. Yet this report is no longer available to us and Congress makes no demands to receive it." [3]

Kevin Phillips in his book *Bad Money: Reckless Finance, Failed Politics and the Global Crisis of American Capitalism* analyzes how the methods the government uses to report common economic indicators like the consumer price index (CPI), unemployment, and the money supply all give a false indication of what most of us are experiencing in the economy. In a follow-up article in *Harper's Magazine* in May 2008 he says:

Readers should ask themselves how much angrier the electorate might be if the media, over the past five years, had been citing 8 percent unemployment (instead of 5 percent), 5 percent inflation (instead of 2 percent), and average annual growth in the 1 percent range (instead of the 3–4 percent range).[4]

What he is suggesting is something that many of us already know, that consumer prices seem to be getting higher than reported. Most mainstream media report "core inflation," which is inflation that does not include food and energy. Apparently they believe that most people using the information have no need to heat and cool their homes, drive to work, or eat.

The chance of a systemic crash is much higher today than it's been for a long time. The chance that inflation will return to the economy in a very aggressive manner is a strong possibility. Let's hope the powers that be can prevent this, but since it's a possibility you should have a plan—a plan to deal with paper dollars having less and less value over time.

Step One – Hard Assets

We've already talked about the first step and that's owning value-producing hard assets. While the monetary value of the wind turbine and solar panels that power your home may increase or decrease with time, the value they provide to you doesn't. It remains constant. The comfort of a hot shower and convenience of food kept cool in a fridge and freezer will continue indefinitely. Hard assets make you more secure. Fifty acres of forested property, a building with a woodstove, a chainsaw and several jerry cans full of gas mean that you can stay warm all winter. A large rainwater catchment system and solar-powered pumps mean that you'll have water for your home and garden even during a drought. A large vegetable garden and tools to tend it, whether they are hand tools or a rototiller, will keep your family fed with fresh, healthy food at very little cost to you. Hard assets should form the backbone of your financial independence.

Step Two - Barter

Step two in moving away from using dollars for exchange is barter. This is what humans have done for eons, but in recent history we have created an intermediary for this exchange called currency. A tradable currency, whether gold and silver coins or paper dollars, gave us more flexibility. We could barter with anyone, anywhere because we didn't necessarily have to be close to them. With the world reserves of oil depleting I believe we are going to see a return to a more local economy, and I think this is a good thing. You need to start thinking about what your current skills are and how you can improve them to barter.

Obviously if you are an auto mechanic you probably have a pretty good set of skills to use for barter. The oil isn't going to run out overnight and some people will still be driving cars. As the economic downturn continues people with cars will be hanging on to them longer and longer, and they'll need to be repaired more often. Some skills will be much easier to barter than others. If you have a strong back and a chain saw you'll probably find people who have standing timber on their property but don't have the time or aren't in good enough shape to cut it themselves.

Chapter 17 discusses the importance of staying healthy, and bartering your services is one of the prime reasons. Our economy has increasingly moved to "information" jobs where we exercise our brains but not our backs. The economy will be returning to more physical jobs as we start to run out of the cheap oil that displaced all this manual labor. The skills that accountants and lawyers and information workers have will be much less in demand than doing real things that make it possible to stay warm and fill an empty stomach.

Step 3 – Non-traditional Tradable Goods

In the "hard landing" part of my workshops I take the demise of the paper dollar to the extreme and suggest that besides bartering your time and skills for goods you should have items that you can trade as well. What form this takes will depend a lot on where you live. One thing that is well known about economic downturns is that the consumption of alcohol increases in relation to the severity of the recession. I would suggest that since this is the worst economic upheaval since the Great Depression and it may in fact eventually be worse, alcohol is going to be a hot commodity. So it would be a good idea to have a fairly good stock of this. What type you have will depend on whose services you expect to need in the future. If you think you'll need lawyers and accountants you may want to make sure you have a well-stocked wine cellar. If you think you'll be needing repairs to your car and a good supply of firewood, you'll probably be more in need of whiskey. Beer is a good option as well, but it'll take up more space per unit of alcohol and may not last as well as wine or hard liquor, which might even increase in quality as it ages. If you're a drinker yourself, having a well-stocked wine cellar isn't a bad idea if you can afford it. Having a lovely meal of vegetables fresh from the garden with a nice glass of wine will make any bad economic reality seem perfectly tolerable.

While most of us know the health risks associated with alcohol and tobacco, many people still smoke and drink as a comfort during stressful times. A supply of these may be a good idea. It's a strategy adopted by many portfolio managers during recessions: they increase holdings in companies with products where consumption is in inverse proportion to the state of the economy. Governments tend to use the same strategy with taxation, often raising "sin" taxes on booze and cigarettes as a way of increasing revenue while decreasing consumption of legal items that are considered to have a detrimental impact on health. It's rare that sin taxes reduce consumption and tend to drive more activity in these products underground.

As times get tougher security becomes a bigger factor in people's lives. Also, some people who have hunted in the past will increase their consumption of wild game that in most cases is free. For both of these reasons ammunition, be it bullets or shotgun shells, will probably be a fairly valuable tradable property. I realize that for someone living in a suburban enclave much of this will seem quite foreign because they don't spend much time with people who hunt. But a percentage of the population does enjoy this activity, and others appreciate the low cost of the meat they get. If you've installed a wood stove in your home as a backup to natural gas you're going to need a source of firewood. If you don't have your own property to cut it from you're going to have to get it from someone who probably lives in a rural area where hunting may be the norm. So if paper dollars are starting to be less attractive as a means of exchange, being able to supplement your offer for the firewood with whiskey and shotgun shells may get you on the delivery route.

Ideally what you want to end up being able to barter is something that is renewable or that you can continue to produce over time. Food will always be in demand, so it's good idea to look at producing the things you grow well in larger quantities. If you find that your soil grows abundant potatoes, then make sure you grow lots of extra. They store well and can form the basis of a sound diet. If you get the hang of keeping chickens, making the chicken coop twice the size you need isn't a lot of extra effort and eggs are a very popular commodity for trade. With chickens, once they're past their laying stage you have a source of meat to eat or trade as well.

Right now many of the items that we used to produce ourselves like clothes and blankets are very inexpensive. They are produced offshore in countries with very low wage rates, and cheap oil allows them to end up on our store shelves at very low cost. The end of cheap oil is going to mean a movement away from globalized trade and a return to more localized trade. So being able to turn cloth into clothing may once again be an option for trade. Quilts and blankets which in the past were often locally made may once again be something people will be willing to barter for. It is difficult to anticipate what your local needs will be, but over time they'll reveal themselves and you'll just have to have a keen eye to spot them. The day of the insanely cheap "Dollar Store" selling items for unbelievably low prices when so much work has gone into them may be coming to an end. The whole premise of this economic model is cheap and abundant energy, and those days are drawing to a close. We will once again be returning to a time when we need to depend on our neighbors for our economic well-being, and they'll need us. I believe most of us would rather know the people we trade with than employ some worker thousands of miles away while neighbors lack paying jobs. A fundamental change like this may be disconcerting but it's not a bad thing.

If you think this view of the future with reduced energy supply and a deteriorating economy is a possibility, I would suggest you start thinking about what you'll be able to trade in the future. The advantage of doing this now is that if you agree that hard liquor may be an excellent item to barter with, then now is a good time to start accumulating it. The price right now is probably much lower than it will be in the future, so it's a good time to stock up. Since you're going to be setting up a pantry for your food stores this will be an incentive to make it larger. Cases of liquor will store well and will be strong enough that you can pile other boxes on top of them containing items you need more often.

Some of the items in your pantry may be very tradable as well, so consider a really big stockpile. Things like toothpaste and matches and soap will be very popular if there are ever disruptions to our supply lines. Toilet paper could become pretty valuable as well. You'll need a large space to store it, but if you've got a spare room somewhere, stocking up on something like toilet paper could turn out to be a pretty good gamble. It lasts forever, so if you can't trade it you can always use it yourself.

Step 4 - Precious Metals

In Chapter 3 I spent a lot of time discussing how much of our current economic mess started when we decided to stop using the gold standard as a basis for our economies. Paper currency backed by a precious metal like gold or silver is the soundest method that humans have discovered for conducting economic activity. Any time you allow governments to introduce fiat currencies that aren't backed by something they will inevitably debase it and it will lose value. So if economies should be backed by gold and silver, is it a good idea for you to own it? Yes! It's a very good idea. In fact, it's one of the most important recommendations of this book. You should own gold and silver and you should own it now.

From 1933 until 1974 it was illegal for Americans to own gold. When Roosevelt took office there was a run on banks and people were withdrawing cash and gold. At that time dollar bills were still certificates which represented gold or silver that was on deposit to back the bill. Technically you could walk into a bank with your paper money and demand that amount in gold or silver. One of Roosevelt's first acts was to declare a "bank holiday" which closed all banks for four days. This was when he made the famous speech in which he said, "The only thing we have to fear is fear itself." In other words, if we all just calm down and stop panicking we'll be fine. When I reopen the banks we're all going to be mellow and we're all going to stop this crazy "run on the banks" and then we'll get through this.

Part of his Emergency Banking Act made it illegal for American citizens to physically own gold. The Federal Government confiscated all gold and if American citizens were caught with gold in their possession they faced a $10,000 fine, which was a huge sum in those days, and five years in prison. The Federal Reserve purchased gold for about $21/ounce and then informed foreign investors that U.S. dollars were backed by gold at the rate of $35/ounce. In 1971 Richard Nixon declared to the world that the U.S. dollar was no longer backed by gold, but he still wanted the dollar used as the "reserve currency" of the world, hoping central banks would keep U.S. dollars in their vaults rather than gold.

So for many decades it was illegal to own gold. But in 1974 the act was repealed and owning gold was once again legal. Owning gold has traditionally been viewed as a hedge against uncertain times or against inflation should a government choose to crank up the printing press and devalue the currency. In the early 1980s the price got up to around $700/ounce and it has fluctuated in the $300 to $400 range for many years. In March of 2008 as Wall Street started to implode the price hit $1,000/ounce and after a brief dip to the $800 range by March of 2009 it was back in the $1,000 range. This may seem outrageously expensive to you but many people argue that this is still actually extremely low. Based on the state of the world economy gold is probably very undervalued.

Gold is so rare a metal that all the gold ever mined could fit into a 20-cubic-yard block. Or as *National Geographic* reported in its January 2009 edition "In all of history, only 161,000 tons of gold have been mined, barely enough to fill two Olympic-size swimming pools"[5]. It's a valuable metal for some industrial processes and has always been in demand for jewelry. It has always retained its value and been a safe haven for wealth invested in it during uncertain times. I hope Part II of this book helped to convince you that we are indeed in uncertain times. So you should be looking for a way to preserve your capital and wealth, and precious metals like gold and silver offer that. Most of us have heard casual reference to precious metals or commodities by financial advisers. They say it should be a small part of a "balanced portfolio" with the bulk of your investments in stocks and bonds. And how has that gone for most people? Not very well, with many people's investment portfolios off by up to half and with little likelihood of that value being restored by the markets in a reasonable time frame.

So what I'm suggesting is that you take that small percentage that you've heard people recommending and make it a much larger part of your investments. I'm suggesting that precious metals offer the potential for great wealth preservation and for capital appreciation if you're looking to increase your wealth. First and foremost I must remind you that I am not a certified financial planner. I am just someone who for 25 years has tried to manage his money effectively. I started in mutual funds. I researched what seemed like good picks but they didn't perform well once I put my money in. Then I read more, attended seminars, bought all the right books, and when I switched my money to those funds that were performing well, guess what? They sat there or tanked. The more I read the more I realized that this strategy doesn't work because markets change and if you switch to a fund that has been performing well it's probably due to level off. This is like switching lines at the bank or switching lanes in traffic. Ultimately you don't end up any further ahead.

Eventually my reading convinced me that more than half of mutual funds do not outperform the markets in which they invest. In other words if your mutual fund invests in stocks in the Dow Jones or S&P 500 Index, 50% of the those funds will not do any better than the market itself. And guess what, you pay lots of fees for the privilege of not outperforming the index. So I started using Exchange Traded Funds or ETFs, which take all the money they're given and buy an equal amount of each stock in the index. So if the Dow Jones goes up by 7% in the year, so will your ETF. The bonus of this is that you don't pay for the privilege of having someone who drives a nice car and owns a huge house do worse than the market. The fees for ETFs are very low.

So for a while I had some money in ETFs and some money in government bonds. Then in 2000 the books I was reading started to warn about the housing bubble. The low interest rates which Alan Greenspan had set to avoid a serious recession were sparking an artificial bubble in housing. The books warned that people who probably shouldn't get mortgages were getting them, and that these NINJA mortgages (No Income, No Job, No Assets) were going to end badly. They suggested that when governments inflate bubbles they do not end well and this one was going to end extremely badly. Everything I read made sense so I got out of the stock market. I still had some money in bonds, but I parked most of my money in my retirement fund in cash. Not a money market fund, but cash. The advisors always said "Put it in a money market fund. It's as safe as cash and it will earn some interest." Well it turns out that it wasn't as safe as cash and it was one of the many financial instruments the U.S. Federal Government had to bail out.

So my money just sat there for more than two years, not doing anything. It wasn't earning interest. And if you factor in inflation I was in fact losing money. This went on for months and the housing bubble got bigger and stock market indexes just kept right along with it going ever higher. Books came out suggesting that we were in a new age of wealth creation and that stock markets might go up forever. Books like Dow 36,000 and Dow 40,000 sold well. There was even a book called Dow 100,000: Fact or Fiction. Wow! That seems too good to be true! I can just invest my money and it will keep growing and I'll get wealthy by not really doing anything.

Well, it was too good to be true. With the Dow trading close to 7,000, or half its all-time high of 14,000, we've had a bitter taste of reality. But my retirement statements are exactly the same today as they were three years ago. Yes, I missed a bit of the upside potential of the bull market, but I didn't lose anything either. And when the losses are as dramatic as they've been, I'm happy to just be where I was three years ago.

I would suggest to you that all those same people who years ago were screaming that we should get out of the market and that this crash wasn't going to be like any others we've experienced in recent times are now all screaming that it's time to own some precious metals. They're suggesting that with the economic mess we're in the likely outcome in terms of paper money is not a good one and it's time for you to start thinking about owning something of real value. If you talk to your investment advisor about this he'll probably suggest you buy shares in companies that mine gold and silver. Remember, those shares trade on stock markets and if the market crashes they may go with it. If you go into your bank to discuss this someone may suggest you buy "gold certificates." This is a piece of paper that says the bank has put aside some gold for you and has it in safekeeping. Remember what I talked about earlier and what happened in the last depression. Gold was confiscated. The gold the bank has on deposit for you somewhere will be the easiest for a government to grab. So I would suggest you stay away from gold certificates. If you purchase gold coins or bullion the bank will suggest that you leave it in your safety deposit box in the bank. While this will probably keep it safe from burglars it will not keep it away from a government that wants to confiscate it.

If you buy gold and silver you should own it. You should have it and you keep it in your possession. There's no use having it and having the government confiscate it. What you're really doing when you invest in gold is putting your money in a place where you think it will hold its value. You should plan on keeping it for a long time. This isn't a short-term investment. This is a long-term move. This is the sort of acquisition that you are purchasing as a nest egg for an uncertain future. If you buy it at $1,000/ounce and it goes to $1,500/ounce you shouldn't sell it. If it goes up that much there is a reason, which is that people are losing confidence in the financial system. They are questioning the value of those paper dollars that the government is flooding the market with. So if it appreciates in value you should be even more determined to hang onto it. This isn't a "short-term make-a-quick-profit-and-get-out" strategy. This is a long-term hedge against uncertain times.

Since the economy started to unravel the demand for gold has risen steadily. Wealthy people have been moving more of their assets into gold. In fact the demand for gold has been so high that it has often outstripped supply and is unavailable. There are now waiting times with retailers who used to always have stock. Based on this demand increase and a restricted supply, the price should probably be much higher than it is. One theory is that central banks and governments manipulate the price of gold because if it goes up too dramatically it indicates to the market that people have lost confidence in the currency. For the U.S. dollar in particular this would have devastating consequences. The U.S. can finance its $11 trillion debt because the U.S. dollar is the reserve currency for the world and other governments need it on reserve in their central banks. If those banks all started to dump U.S. dollars and replace them with Euros or gold, the value of the U.S. dollar would drop and this would have a very negative impact on the U.S.

On the United States Mint website when you try and order American Eagle Gold Proof Coins as of March 2009 you get this message:

> Production of United States Mint American Eagle Gold Proof and Uncirculated Coins has been temporarily suspended because of unprecedented demand for American Eagle Gold Bullion Coins. Currently, all available 22-karat gold blanks are being allocated to the American Eagle Gold Bullion Coin Program, as the United States Mint is required by Public Law 99-185 to produce these coins 'in quantities sufficient to meet public demand . . . .' The United States Mint will resume the American Eagle Gold Proof and Uncirculated Coin Programs once sufficient inventories of gold bullion blanks can be acquired to meet market demand for all three American Eagle Gold Coin products. [6]

So apparently no one is trying to hide the fact that people are flocking to gold right now.

So in order to avoid the perception that paper dollars are weak, central banks can simply sell off some of their gold when demand increases to keep the lid on. Eventually, though, if the demand is high enough and there is enough pressure on all fronts, it will be difficult for governments to keep up the fight. And if this happens the price of gold could rise dramatically while the value of the dollar drops precipitously. I have to return to my earlier comments that this is not a scenario I hope for. In fact I dread it. Life for many people in the developed world is very good right now. I have had incredible opportunities in my life since I entered the workforce in the 1970s and I want the same for my daughters who are just entering it now. But what I want to happen and what I think is likely to happen are not the same.

If I'm wrong, which I hope I am, owning gold is not a bad thing. Over time it will hold its value and it will be something for you to pass on to your heirs. If at some point in the future the economy seems to be returning to some sense of normalcy and it looks as if governments can handle the converging challenges we face today, then you'll always be able to sell it. If you sell it and take a loss that's a good thing because it means that things are returning to normal. But if you own it and it doubles or triples in price then there will be even more incentive to hang on to it for as long as you can.

How to Buy Precious Metals

Gold and silver comes in two main forms, bullion or coins. Coins are often referred to as "rounds." While coins may be more familiar to most of us I prefer to buy bullion in wafers or bars. A coin will often have a nominal "face value" such as the U.S. Eagle Gold Coin which has a face value of "Twenty Dollars" engraved on it. The Canadian Gold Maple Leaf Coin has a face value of "50 Dollars." This means the coins have a "legal tender" value of $50, but the numbers are largely symbolic because they don't represent the market value of the coin. So if you go into your local store to buy groceries the store is required to accept that coin as being worth $50. But that's ridiculous since it cost you $1,000 to purchase the coin. So the face value or legal tender value of a gold or silver coin is irrelevant and confusing. If in the future you want to use that coin to purchase something, will the person you're completing the transaction with use the current price of gold, which could be thousands of dollars an ounce, or would they want to use the legal tender value of the coin?

The advantage of wafers or bars is that there can be no confusion about exactly what they are. In the picture you can see a 1 oz. gold wafer which is identified as 1 oz. of .9999 fine gold from the Royal Canadian Mint. There is no nominal face value. This is an ounce of gold and nothing else. If you want to sell your coins back to a bank in the future they will be valued according to the precious metal content. But if you are negotiating a financial transaction with your dentist there may be some debate about the value of the coin, so it is best to buy wafers if you can.

Coins are a bit easier to purchase and their disadvantages compared to wafers are minimal, so if the choice is coins or nothing, take the coins. You'll also have to decide if you want gold or silver. In my dreams I have rooms full of gold coins, just like Johnny Depp in the movie *Pirates of the Caribbean*. In real life I have never made a great deal of money and do not have much in the way of financial resources, so what precious metals I do have are silver. While an ounce of gold is worth about $1,000 today, silver is worth about $15 an ounce. Which means I can buy more. Gold is without question the premium and preferred precious metal to own, but with limited funds I've had to resort to the poor man's gold and very little of it. My choice has been to put money into solar panels and wind turbines because they make living off the grid much easier. I hope to purchase more precious metals in the future depending on my finances.

The other option I'll mention is older and collectible coins. Historic coins offer proof that coins containing precious metals maintain their value over time. The one type of gold that Americans could keep when Roosevelt was confiscating gold was historic coins because there was no way to properly value them. Also, since it was conceivable that some of the gold the government confiscated might be melted down to produce larger gold bars, the use of coins was viewed as destroying historical artifacts. From that perspective these coins do offer some benefits. The challenge is that buying them is a very specialized skill and not something the average person should undertake. If you are interested in this you need to research it thoroughly and read everything you can on the subject. Then you'll need to find a dealer you trust to help you as you pursue this option.

So your choice is based on how much money you have and which you prefer. Both gold and silver have actual industrial uses. Silver is used in electronics as a catalyst in chemical reactions, in mirrors, and in silverware for the table. At one time it was used in dental fillings and in photographic film. Gold can also be used in electronics and jewelry, so which of the two will be more in demand for industrial purchases will change over time. The benefit of silver for many of us is that it is less expensive and we can buy more of it for the same amount of money. The relative price difference between gold and silver has remained fairly consistent over time.

The other benefit of silver is its ease of use in financial transactions. If in the future the value of paper dollars becomes questionable and you use your gold and silver for transactions, the lower value of silver will make it easier to use from day to day. If you need some dental work done, a dentist accustomed to charging $1,000 for the procedure will probably be comfortable with a one-ounce gold coin. If on the other hand you want to purchase 20 pounds of flour from the local organic farmer which she values at $100, a $1,000 gold coin is going to be difficult to use. Do you try and cut one-tenth off the coin to give to her? A silver coin worth $15 or $20 is going to be much easier to use. So having a mix of gold and silver if you can afford it is a good idea.

Where to Buy Precious Metals

There are lots of places to purchase gold and silver, from banks to coin dealers to websites. One thing I recommend is that if you are buying precious metals as a long-term store of wealth and potentially a hedge against very challenging economic times, try to purchase with the minimal number of audit trails. Essentially try and purchase precious metals for cash with no receipts. You walk into the store or coin dealer, pay cash (dollar bills not plastic) and you walk out with the gold and silver. I suggest this simply from a capital preservation standpoint. If the government decides it wants to get the gold back, it's better if it doesn't know you have it. If you purchase it from a bank that fills out an order form, or online with a credit card, there is going to be a record of that transaction. There is absolutely nothing illegal about taking dollar bills to a coin dealer and purchasing precious metals. I suggest this is the way you build your precious metal reserve.

If you can find a local coin dealer who will work this way that's excellent. Get to know him, but on a first-name-only basis. He may be able to guide you through the process and give you additional insights into the various products available. He may also have access to private sources for these items rather than having to purchase them directly from the Mint. He may let you know when he's purchasing a collection from someone so you can take a few days to get some cash together. Building a collection of gold and silver should take time because it's not always the best idea to walk around with a huge amount of cash.

Try the yellow pages under coin dealers. You can also buy your coins directly from the United States Mint at www.usmint.gov or the Royal Canadian Mint at www.mint.ca. Obviously there will be a very noticeable audit trail to your purchases there. The U.S. Mint website does have a feature that allows you to find local dealers in your area (www.usmint. gov/bullionretailer). There are also independent online dealers like Border Gold (www.bordergold.com), Kitco (www.kitco.com), USA Gold (www. usagold.com), and Goldline (www.goldline.com). Companies change over time and offer different terms and buying strategies, so do some research to find a source you are comfortable with.

How to Store Precious Metals

The first thing people think of with items like this is that they should put them in a safety deposit box. That traditionally has been sound advice but may not be appropriate for times like these. The question is whether, with the severity of the economic downturn and the Patriot Act's new powers of personal intrusion, it makes sense to have an asset you've purchased for a worst-case scenario in a place where you may not be able to get to it. I would suggest this is not a sound place for your gold coins. I would suggest you store them in your home.

A safe would be a good place to start. Some would suggest you'll need two safes. The first, where you keep a bit of cash, is the one you take the home invaders to. The second, better-hidden safe is where you keep the bulk of your assets. Not owning much silver it's not much of an issue for me, but I prefer to use the camouflage technique. On average burglars will spend eight minutes in a home. They're going to go to all the easy places like the drawer in the table nearest the door where the man stores his wallet, then upstairs to the top drawers of the dressers, then to the jewelry boxes on top of the dressers, then the night tables, that sort of routine. What you need to do is disguise what you don't want them to find. Still have a big collection of VHS movies? How about taking that *Gone with the Wind* tape—well not that one, that's a good movie. How about that movie you always regret buying because it was a train wreck? Put your silver in there. If a burglar is zipping around the house looking for valuables, VHS movies don't exactly scream "I'm valuable," and if you have a wall full of them who's going to bother to take the time to check?

There are lots of commercial products you can buy for this camouflage strategy: pop cans, flower pots, false electrical outlets. You may want to check them out and see if anything twigs. While it might be tough to make your own mini pop-can safe, next time you open a canister of potato chips you might want to peel the foil protector on the top back carefully so that once you've eaten the chips you can put items in the container and put it on a crowded shelf.

In my workshops I use a photo entitled "Spot the Silver Bars"; it shows a messy, chaotic paint shelf in the garage, which is a bit of disaster. Humor is important when discussing issues like this and my lack of garage tidiness skills always get a laugh. I kid that there's someone in the back row text messaging their cousin to get over to my house because I've got silver stored in old paint cans. I come clean that I don't have much silver and what I do have is not stored there. The example is simply to show the value of camouflage.

A recent magazine article traced a couple's attempts to bury valuables in their backyard. I know, this sounds crazy, but I'm throwing it out there anyway. Humans have been doing this for centuries so apparently there is some merit to it. If you live in an urban area you need to do this discreetly, preferably very late at night. If you saw the Alfred Hitchcock movie *Rear Window* you'll know it's important to make sure some Jimmy Stewart-like neighbor isn't observing your late-night activities in the garden. If you live in a cold climate remember that once you get a good frost in the ground it will be difficult to dig your stash up, so anything you bury should be for the long term. If you live in Minnesota and decide it's finally time for a big-screen TV just before the Super Bowl in January then a backyard wealth deposit is probably the perfect antidote, because without a large propane torch to heat through the frost that stash is staying buried. This is an excellent method to prevent you from being tempted to use the stash unless it's a real emergency. I extend my apologies to my (male?) readers who would argue that a new widescreen TV for the Superbowl does in fact constitute an emergency.

I would not suggest you bury cash. It is next to impossible to keep it dry and it will deteriorate over time. Bury the gold and silver. Since moisture is the enemy having the container as airtight as possible is critical. The photograph shows an easy-to-make and relatively inexpensive container. You take a length of IPEX 75mm rigid plastic plumbing

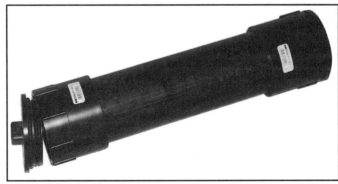

pipe and use two of the screw-on caps that a plumber would use for a cleanout. A plumber would use only one. You use Y2 ABS Solvent Cement to glue the cleanouts on each end. Once they've dried you can put some Teflon tape on the thread of the screw-on cap to help make it extremely airtight. Use a large wrench to tighten the cap firmly once you've placed your valuable inside. Then it's time to head out to the yard after *The Late*

Late Show is over and quietly dig a nice deep hole to put it in.

Remember that you should tell someone about your little buried treasure, like a parent or child, someone who doesn't accompany you on trips away. That way if you meet a sudden demise they'll be able to recover it. Hopefully you have a good enough relationship with your parents or children that the next time you're away for the weekend you won't come back to find that someone's been digging in your backyard. You need to evaluate your own family dynamic critically before you make this decision.

I understand that many readers right now are shaking their heads and saying "this guy is wacko." I understand and accept your criticism. I seem very outside the mainstream. Or at least I seem very outside the mainstream today. With an historic look at things though, my approach is in keeping with the way humans have dealt with trying to store wealth over the centuries. All this talk of gold probably seems very strange to you. You're conditioned to hearing that the people buying gold are the same ones who are building bunkers in their backyards and stocking up on shotgun shells and camouflage gear. I know it seems strange to someone raised using paper dollars. It must seem even stranger to someone who's quite young and has hardly ever used cash at all and has relied on electronic transactions with credit and debit cards. It is very much in keeping with my theme, though, of returning to the values of your grandparents and the independent way they lived. In the 1920s and early 1930s they could go to the bank and withdraw gold for transactions. It wasn't necessarily convenient, but it was an option. Like my recommendation that you return to a time of energy and food independence, having gold and silver coins around is just another step on that road to personal independence.

I would suggest that precious metals have historically been a sound method of wealth preservation. It has also been an excellent means of self-preservation for many in historically dangerous times. As the Nazis rose to power in Germany in the 1930s it became clear to some Jews that this was not going to be a good place to stay over the long term. Many emigrated early on while it was still possible to do so legally. Those who waited until later in the decade found it increasingly difficult to leave as anti-Semitism rose to a fever pitch and the movement of Jews was restricted. Those who did get out were able to bribe their way out, and with the German currency having been severely inflated during the decade the best way to bribe someone was with something that maintained its value—gold and silver. Gold and silver coins and jewelry were what allowed some to escape. I certainly hope it will never come to that here, but since owning some precious metals is part of a sound financial portfolio, it's nice to know that if you have them in your possession they can save you in an emergency.

Once you get over that "survivalist" mentality about precious metals and realize they are a part of any household's financial assets, they'll suddenly seem less strange and more comforting.

I'll add just one final note on this topic and suggest that you may want to keep a bit more cash on hand than you do now. I realize I've been arguing that inflation erodes the value of paper money, but it still is the main medium of exchange in our economy. Debit and credit cards have allowed many us to function perfectly well in the economy without carrying any cash. This will still be the case as long as the power stays on. Earlier on I spoke of the challenges facing our power grids and how the lack of investment in the electrical infrastructure has increased the likelihood of power disruptions. While you're going to be putting up solar panels with a battery backup to power your home, you may still want or need to purchase things during a blackout. If the power goes down and you're out and about and have no cash in your wallet you are not going to be able to purchase anything. I recommend that you always keep your gas tank at least half full, and anyone caught in the traffic nightmare of trying to evacuate coastal areas during hurricane watches knows you should be starting with a full tank and filling it when it hits halfway. But if you were a distance from home and your gas was low and the power went out, if you could find a gas station with a backup generator to pump gas it sure would be nice to be able to buy it. Cash—dollar bills—might help.

Some would recommend that you have several months' worth of cash on hand. Once you've calculated your budget you can decide on an amount. I'm not really contradicting what I've said about precious metals. If there is a "bank holiday" declared, people will still be completing transactions with cash. There will be an interval before people realize that dollars have lost their value and they switch to gold or silver. So if there is a period of dislocation, having lots of cash will help. If you need prescription medications or your child needs an asthma inhaler you can use that cash reserve to stock up. If things in the economy are unsettling to retailers they will be much less enthused about electronic payment and much happier to have cash. That moves you to the front of the line.

If you needed to stay in town overnight because the ice storm that caused the blackout made driving treacherous, it would be nice to be able to pay for a room in a hotel or motel. Yes, these are unlikely scenarios but they could happen and not having $50 or $100 cash in your wallet could turn an inconvenient situation into a real mess. Keep some cash in your wallet. Keep some cash at home. I don't believe governments will have to declare "bank holidays" and restrict your access to your money, but if they do you'll appreciate having some cash on hand. When the FDIC takes over a bank you don't want to be one of those stressed people standing in line in the heat as they did at IndyMac trying to get out some cash. Take it out now. Put it in your pop-can hiding place and leave it there. Don't take it out. Don't use it for the pizza man when you don't have cash. You shouldn't have ordered the pizza if you didn't have the money at hand. In fact, you really should be making your own and taking the $20 you saved and putting it towards your mortgage. But you've heard all that by now and you have a new commitment to get debt free, which is the key to your new independence!

21 Living Happily in the New World!

Arguably the only goods people need these days are food and happiness.
Sir Terence Conran

*Every possession and every happiness is but lent by chance for an
uncertain time, and may therefore be demanded back the next hour.*
Arthur Schopenhauer

*The secret to true happiness is a combination of low expectations
and insensitivity.*
Olivia Goldsmith

*What can be added to the happiness of a man who is in health,
out of debt, and has a clear conscience?*
Adam Smith

Think back to a time in your life when you were really happy. Maybe it was playing with some friends as a kid, out on your bikes or at the park. Maybe it was swimming in a pond or a lake on a hot day. Your first long walk with someone you really liked. For many of us, great memories aren't necessarily those that involve money. They involve non-monetary things: friends, places, experiences, just being somewhere.

Now if all you can come up with is events related to money—your first house, your first car, your first paycheck—you've got to try and think of some other things you've enjoyed that didn't entail money. This is really important because you have to try and break that very strong link in your mind between money and happiness.

If you were fortunate enough to grow up in a stable environment as a child, think back to some time that you were enjoying yourself. This is hard, because we're all prewired to much more easily remember negative experiences. Getting scolded by your parents, a bad mark on an assignment at school, being picked on by a bully. It's only natural and in some instances a survival instinct for these memories to be stronger than happy ones. When an ancient human was chased by a saber-toothed tiger it was important to remember to take evasive action the next time she saw one in the distance.

I think if you try really hard you're going to remember some good times that involved little or no money. I know that when we got married in the early 1980s, at a time of relatively cheap gasoline, Michelle and I drove out west for the summer for our honeymoon. I had dropped out of university and Michelle hadn't found a full-time job, so we got married in June, jumped in our tiny Toyota Tercel, which got insanely great gas mileage (and wouldn't pass any crash test today), and drove from Ontario across Canada to British Columbia and then down to California. We stayed with family in Edmonton and Vancouver. We camped most nights in provincial parks, which were very cheap in those days before the neo-conservative age of cost cutting reduced money to public parks. The odd time after a number of days of rain we might find a very cheap hotel to dry out in. We cooked our own meals, stayed out of tourist places, and even camped for awhile on a free beach in California. We probably spent around $1,000 the whole summer and it was fantastic. Every day was a new adventure—driving somewhere we'd never been, seeing things we'd never seen—and we hardly had a penny to our names.

We certainly had no security. We had no jobs to come back to. We had very few possessions and most of our wedding gifts had been coolers and camping equipment for the trip. But regardless of how little money we had, we've never been happier. Every day was an adventure and our biggest challenge was figuring out the cheapest way to eat that day. Twenty-five years after the trip I can still remember many of the experiences as if they just happened. The sounds, the smells, the feelings. Lying in the tent with the flap open first thing in the morning. The anticipation of what was around the next corner.

Have you had an experience like this? Maybe during the last few years you've been living in a large house with lots of rooms and cars and the expenses that go along with them. Yet when you think back to that first house you bought, or that first apartment you rented, the one where you used old wire spools for coffee tables and made bookshelves out of discarded wood and bricks, you probably have some pretty good memories.

Now you have to reach back to those times to remind yourself that your happiness isn't necessarily a product of your income. It's not related to how big your house is or what kind of car you drive. If your work and living and driving arrangements have changed or are in transition, perhaps to a more scaled-back level, this is not a bad thing. In fact, it could be a good thing. It could be a really great thing if you think about it in the right light. It's up to you. You can sit around feeling sorry for yourself, missing that 4-bedroom, 3-bathroom, 3,000 square-foot monster home in suburbia, or you can start looking at the upside of your new living arrangements. Smaller places require much less cleaning. They're easier to heat and cool. You don't have to work so hard to pay for them. If you're back renting as the housing market declines, you'll be able to buy in with much saner prices in the future. This is a good thing.

And what about your job? Do you remember one of your earliest jobs? Helping your parents in the kitchen or spring cleaning the yard? Remember how nice it felt to have cleaned it up and how good it looked? Or your first paying job? Even if it was flipping hamburgers there was the challenge of learning the process and the joy of that first paycheck, regardless how small. Sure, once you got into the rhythm of the job it was less inspiring. Maybe you daydreamed during mundane and repetitive tasks, but there was probably much less stress than there is in your more recent, higher-paying, technologically inundated, timeline-dependent, high-stress job.

So it's time to accept that your new work will pay less but will be less stressful. If you're lucky, it will demand more physical labor on your part. You know how sick you were getting of sitting at a desk all day. And how your butt and your back ached after nine hours in that chair. Your ears were sore from that phone headset. Your wrists hurt from typing on that keyboard all day. This new job is going to take some getting used to, but you have to approach it with the right attitude. It's much better for your health than the old one.

Now you have to make the leap back to that first paying job. It wasn't much, but it really meant something to you. As you worked more, got new jobs, worked your way up the corporate ladder you were making more money, but it was becoming less fun. And with every raise in your salary came a new purchase to offset it. A bigger house. A newer car. That cottage on the lake. More trips and holidays and all that stress of airports.

So if you're going to survive this new economic reality you're going to have to change your attitude and take yourself back to those good old days. Your plans and goals were much different then. It wasn't about how much money you were going to squirrel away for your retirement, it was all about looking forward to getting your paycheck and going out with friends on Friday night. This is the new thinking you've got to get your head around. You've got to look much more short term and less long term. You're going to be putting more of that money to productive use now by investing in hard assets that help you achieve long term independence.

Buddhism talks about the concept of "mindfulness" or being aware of one's thoughts, actions, and motivations. For people suffering from depression it can be used as a strategy to help them try and remove themselves from the sources of their stress and become very focused on the here and now. For many the stress or depression comes from bad experiences in the past or fear of future unknowns. By separating yourself from those things that do not impact you at this very moment you can try to start bringing some positive light back into your life.

You have to start "living in the moment". This just means you have to tell yourself that what you're doing right now—having a coffee with a friend, weeding your vegetable garden, loading firewood for next winter, cutting vegetables from your garden to make soup— these are good things. Right now, I'm happy. I will not dwell on what might have been if the economy had continued as it had for so many years before I lost my job. I will not worry about what's going to happen months from now. I can't change the past and I only have some control of the future. But I can tell myself that what I'm doing right now is a great thing. I'm taking joy from this everyday activity.

Have you ever gone camping? Car camping is great, but you tend to pack a lot of stuff and you can always take off and get something if you need it. Have you ever gone on a long backpacking or canoe trip? When we lived in the city I used to go canoe tripping. This was a great escape from the city, which I didn't like. Trips were generally three or four days. You would put in on one lake and then paddle a big circuitous route and end up back where you started. You would have to portage from lake to lake, so you'd have to pack lightly because you'd be carrying everything you needed for four days on your back, no excess baggage. I would prepare for weeks. Scope out every meal. Every snack. Every change of clothes. And prepare for bad weather and good.

There was always this pivotal defining moment on these trips and that was just as I was about to set off and I'd put my wallet in the glove compartment and lock the car. My wallet would be of no use to me for the next four days and was just extra weight. That's so foreign to most of us. You can't function in our society without a wallet. You need money for everything. You would never think of setting out on a shopping trip without your purse or wallet. But there was something so liberating, so freeing about leaving your wallet behind for four days because it was going to be of no value. Yes, I needed it to buy the food I was going to eat and the gas I used to drive to the northern lake, but in the bush money was useless.

We live in a society where you need money for just about everything. The goal of this book has been to help you start living more and more as though you're on a canoe trip. When you lock your wallet away for extended periods of time because you don't need it. Because you grow your own food, make your own electricity, and generate your own transportation by converting your food energy into motion on your bike. The more elements of your life you can free from the money requirement, the less burdened you'll be and the less stress money will cause you.

I just read a magazine article about oil workers who have become unemployed as the price of oil has dropped. Some of them were making $10,000 a month. I make just a little bit more than that a year. The scary thing is that they have monstrous debt and now with no job they're contemplating bankruptcy. This is not a good financial strategy. Somehow humans have lost the concept of saving for a rainy day. We all just proceeded as if the good times would continue forever without ever thinking that the economic boom might go bust. This has jolted most of the world back to reality. I'm hoping you're taking it to the next level and saying it's not enough to have some savings for a rainy day. It's now time to set a course for a lifestyle that requires a minimum of income so that you can weather not only rainy days but also rainy seasons. So that you can handle monsoons and downpours and ice storms and endless rainy months. In this day and age this is the only sound financial advice anyone should be following.

Work and spend just doesn't cut it anymore. Our households have been "money in and money out," and sometimes more money goes out than comes in. Sometimes a lot more goes out. It's time to realize that is a financial strategy fated to end in tears.

Many years ago I read a book by Joe Dominguez and Vicki Robin with Monique Tilford called *Your Money or Your Life: Transforming Your Relationship with Money and Achieving Financial Independence.*[1]   It helped me start to form a new relationship with money. The authors have a great concept called "The Fulfillment Curve." It charts the relationship between the money you spend and the fulfillment you gain from it. As you start spending money you're addressing your basic human needs for survival, like food. As you continue up the curve you get comforts like some warm clothes and a roof over your head. After that the money you spend is directed towards luxuries because after you've covered survival and

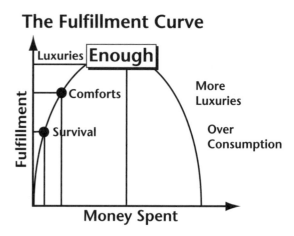

comfort everything else is a luxury. You do need shoes but a pair of $100 running shoes is a little more than you probably require. Many people consider a car a necessity, but since you could live closer to work or ride or walk or take the bus, it really is a luxury. We just forget what a luxury it is. While food is a necessity, coffee is a luxury. We don't really need it. It just tastes good and we enjoy it, and considering how far it travels to get into our cup it's an incredible luxury.

If we start thinking of so many of the things we use and experience as luxuries, suddenly our life starts looking pretty amazing. The real challenge is when you hit the top of the chart, which is "enough." Once you go past "enough" you are into the over-consumption zone where you keep spending money but the things you spend them on are not bringing you any more fulfillment. I think economists call this the diminishing marginal utility of each new purchase. During the last couple of decades we've all been on a bit of an over- consumption bender. We've been buying too much stuff and spending way too much money on everything from vehicles to houses to travel to restaurants to services, you name it. And the real problem is that many of us really can't even afford these excess luxuries. Our incomes don't justify the expenditure, but those amazing pieces of plastic in our wallets facilitate it anyway. Now many of us are waking up with one heck of an over-consumption hangover and are feeling pretty dazed and confused. As the credit card bills fill up the mailbox each month, though, we're slapped back into reality very quickly.

So you have to now redefine your life and set a new benchmark for what "enough" is. It will be lower than it used to be. It has to be because you can't afford to go on spending money you don't have, and the money you do have you are going to apply towards paying down debt and increasing savings. There will still be purchases. If electricity is cheap in your area then perhaps you could look at solar panels as being a luxury you don't need. But in a future where utilities are increasingly stretched and unable to maintain the same level of service and reliability, an independent power system for your home I think is a necessity. A luxury is a trip south in the winter. You may be used to it, but you need to save that money now. A luxury is buying broccoli in July, because you should have a garden full of it. A luxury is takeout pizza because you can make a far better one yourself for a fraction of the cost. A luxury is a $4 cup of takeout coffee when you can make a perfectly fine one yourself. Go down the list. Look at what you spend your money on every week. Got a whack of credit card debt? Then 200 channels from your cable company is a luxury you can't afford. Cancel it and start using that time to figure out other ways to save money. All the kids want their own TV in their rooms? No way. That is ludicrous. The local car dealer has unbelievably great deals on new cars? Forget it. Your seven-year-old car is just fine. If it's ten years old then sign up with the automobile club and leave yourself extra time for the days when the car won't start or needs to be towed.

Make your own list. What's your priority? Keeping up with the neighbors or paying off your credit cards? One week of sunshine in Costa Rica in February or a lifetime of freedom from the stress of having your home repossessed, which eliminating debt and paying off your mortgage will provide? Arriving at family gatherings with a late-model car to prove how successful you are or showing up in a beater and living in a tiny home with a huge garden that you own outright? I know what my priority has always been and it's never involved how I'm perceived by others. It's always been about being debt free and being energy and food independent. I don't think you can really put a price tag on that. But once I got here it was easier for me to define "enough." My world off the electricity grid is powered by a finite amount of energy and I live within the bounds that nature sets for me. The planet has limits. My income has limits. I live within my limits. I do not go beyond the limits because when you do that in nature you get unintended consequences. And when you do it with your checkbook you run up against those same limits.

Studies completed since the 1950s prove there is no link between income and happiness. Since 1950, as income has exploded and material wealth has gone up exponentially, the number of people who consider themselves "very happy" has gone down steadily. We now have almost four times the material wealth we had in 1950, bigger houses, more income, and more multiple storage units for our "stuff," but we aren't any happier. Ask a person who has traveled the world who seems happier, North Americans or people with significantly less than we have.

Do you know anyone who takes drugs for high blood pressure, cholesterol, asthma, or headaches? According to a government study, antidepressants have become the most commonly prescribed drugs in the United States. They're prescribed more than drugs to treat high blood pressure, high cholesterol, asthma, or headaches. How can this be? We have it so good. We have so much! Yes, it's a cliché, but money doesn't buy happiness. If you've been forced by financial circumstances to live with less than you're used to, you're still doing well compared to the standard of living of many people on the planet. And if you're about to start on a journey towards independence you're ultimately going to be able to live on less. You'll have so many of the basics covered you simply will not need as high an income to get by. This means less work, less unproductive time spent away from doing what you enjoy, less energy sucked out of your soul at a desk.

What I'm describing here is a change in how you see the world. I'm requiring you to change your perception of reality to one with a very different future. At one time you might have considered the future I'm suggesting as one of denial. Soon, though, you'll see this world of reduced expectations and living within your means as cause for celebration. Knowing that the earth has limitations and you have limitations and committing to live with some constraint is going to help you start being happy with what you have. You're going to stop longing for bigger and better because that's what got us all into this financial mess in the first place.

Did you know that you live better than 99% of all the people who have ever lived on this planet? I hear you: "No way, I don't live like a king or a queen! Not even close." All right, let's look at this a little closer. Now let's say you were a king, perhaps King Henry VIII who ruled in England in the early 1500s. Let's talk about heat. Have you ever tried to heat a castle? What a hassle! Fireplaces in every room. You'd basically need a full-time staff to cut the trees and keep the fires going, and you'd still never be warm in a castle because those inefficient fireplaces would be sucking all the warm air out of the room. In your home when you get cold you walk over to that magical little box on the wall and turn the thermostat up, and instantly fossil fuels are converted to heat. If you want to be crazy you can crank it up until it's as hot as summer. Speaking of summer, let's think of Queen Elizabeth I, who ruled England later in the 1500s. Have you seen those pictures of her with the high collars and the tight-waisted dresses and huge crinoline skirts? Can you imagine what it would be like to wear something like that on a humid day? Gross! We get hot and we head over to that magic box again and turn up the air conditioning, which removes much of that humidity.

Now let's say Queen Elizabeth wanted to visit Spain, to form a new alliance. She'd need a few weeks to prepare, a week or two to get all that gear to the coast to take a boat across the English Channel, then another week or two to travel to Spain by horse and carriage. A week there to recover from the journey, and then she'd have to repeat it to get home. That adds up to four, five, six weeks she's invested to hop across the channel. If you want to go to Costa Rica for four days you can book a last-minute flight and be there in a few hours! The mind boggles at what a luxury that is. When you're in Costa Rica you'll probably enjoy lots of fruit. The pineapples there are amazing. But we don't have to go to Costa Rica to enjoy them. We can pick up a fresh Costa Rican pineapple at our local grocery store! Sometimes for $1.99 on sale! How do you grow and harvest and ship a pineapple to my store in a cold climate for that price? It hurts your brain to even think about it. Now Queen Elizabeth probably never had a pineapple, because Elizabethans couldn't have gotten them across the ocean on a sailboat before they went bad. And we can eat fresh fruit year round, from all over the world. Some of it arrives by jet from warmer climates.

What if Elizabethans got sick? There were no antibiotics. There were no emergency rooms. As I recall, Henry VIII was a pretty big guy. I'll bet he had high blood pressure. There was no heart surgery for someone like him. So have I convinced you yet? Humble and non-rich you are living better than two of the richest and most powerful royalty who ever ruled England! Not bad! Life really is pretty good when you think about it. You just have to start appreciating everything you do have that makes your life so great.

Put your own list together. ...iPods, DVDs, electric lights, swimming pools, hair dryers, CD players, high-speed Internet, power tools, lawn mowers, refrigerators, summer camps, Hollywood movies, Dollar Stores, shopping malls, Advil, orthodontists, laser eye surgery, solar panels, hot tubs, coffee shops, cottages, gas stoves, freezers, Lazy Boys, life insurance, elections…. The list is endless and yet some of us don't appreciate just how good we've got it.

One of the keys to your happiness in a radically different future is going to be your ability to take more time to look critically at all the good things you have to be grateful for and appreciate the time of abundance in which we find ourselves. You may have lost your house and you may be living in a much smaller rental home but look at the luxuries your home has. It has central heat. You sleep on a wonderfully supported bed. You have wonderful fabrics that make your clothes light and comfortable. You have toothpaste to keep your teeth from rotting. You turn on a tap and water comes out, as opposed to having to haul water in buckets from a well. You have a toilet that flushes wastes away to help you stay healthy. You can stand midwinter in a shower while jets of hot water cascade down on you and you can use soaps and shampoos to keep clean and feel good about yourself. Let me know when you want me to stop. Compare your reality to that of Queen Elizabeth and you'll find you're doing very well for yourself. Compare yourself to someone living in a developing country with no modern infrastructure and you'll agree that you're living beyond the wildest expectations of a huge percentage of the people who live on the planet today. Almost two billion people live without electricity. Billions live without proper water and sanitation. Some people on the planet earn the equivalent of $1 or $2 a day. You are indeed lucky and blessed.

Gregg Easterbrook wrote a wonderful book called *The Progress Paradox: How Life Gets Better While People Feel Worse*. It documents the disconnect between our rising incomes and high standard of living and the extent of our happiness. He has two recommendations for remaining happy. One is to surround yourself with positive people.[2] This may be difficult if you have fallen on hard times, but you really need to find people who are positive and spend time with them. Being around positive people will keep you up. Whether it's in your church, on your baseball team, at the community garden or the local diner, be around positive people.

His second key to happiness is to "be grateful" or what I've heard of as an attitude of gratitude. Now that you know just how great you've got it, you need to acknowledge this constantly. I'm having a pineapple for breakfast that came all the way from Costa Rica! I'm very grateful to live in a time of such abundance that little old me can afford a pineapple! In

February yet! And it's chilly today. I've just turned up the thermostat and soon I'll be warm. How great is that! And right now I'm sitting in my warm living room having a cup of tea, which was grown in India, with a friend I've known since high school. Friends are great support during this challenging time. I am very grateful to have such a great friend and to have the electricity that boiled the water that I steeped the tea in. Once you get those solar panels on your roof and you understand what a miracle electricity really is, you'll never turn on another light switch or make a piece of toast and not be grateful!

The key to you remaining positive in these challenging times is who you compare yourself to. If you continue to judge yourself against the yardstick of those people with higher incomes and bigger homes, much of which you see on television and is not reality, you're going to be miserable. If you compare yourself to people who live in the Lower Ninth Ward of New Orleans your conditions are going to start looking pretty good. Now compare yourself to someone living in Sub-Saharan Africa and you're living like Bill Gates. It's all relative. It's about what standard you compare yourself to.

In the intro to Part III on page 61 we spoke about Maslow's hierarchy of needs, where he ranks human needs starting with the most basic and essential and moving to those on a higher level. At the bottom are the real basics like food, water, and sleep. Next you need some safety in many facets of your life. As you work your way up the pyramid the needs become less basic: love and belonging through family and friendship; self-esteem and achievement; "self-actualization," which really is the ultimate achievement of a well-rounded person. The whole self-help world focuses on these upper layers of the need hierarchy. People today are free to pursue these less concrete needs because cheap and abundant energy has allowed all the lower-level needs to be taken care of easily. We spend a little over 10% of our income on food today. It's a very small percentage, something we should be infinitely grateful for. This allows us to have lots of income left over to pursue the rest of the hierarchy.

This one-time gift of fossil-fuel energy that we have been burning through in the last 100 years is starting to show signs of dwindling, which means that we are going to start spending more of our income on basics. In fact as 6 ½ billion people compete for less and less food I believe that food shortages are a possibility and that we'll all be even more focused on food. As we start to have price spikes for the fuels that we use to heat our homes we're also going to become more focused on shelter and will be devoting an increasing amount of our incomes to that. So it looks as though much of the attention which we previously focused on needs higher on the hierarchy is going to be focused on the lower, more basic needs.

The thing I sincerely believe is that by focusing more on basic needs we will also be addressing many of those higher needs, but as an outcome of the lower pursuits rather than as a goal unto themselves. In other words, we'll be spending less time and money going to hear self-help gurus tell us how to become self-actualized, but we will be discovering a whole new meaning to our lives. We're going to become more dependent on our friends and neighbors during challenging times. Once we get that ground-source heat pump or wood stove installed and we know how we're going to heat our homes next winter we're going to experience a huge boost to our self-esteem and self-confidence. Installing that solar hot water heater and putting in that vegetable garden are going to provide a huge sense of achievement. In fact I would suggest you'll be spending a huge percentage of your time just standing and staring at that garden. It's a beautiful thing. It feeds your family. It's wondrous that these tiny seeds grow into plants that provide sustenance for your body. I would suggest that when you cut that first broccoli you'll stand there staring at it as you never did in the grocery store. The broccoli in the grocery store was grown by someone else, but this one is special. That's what self-actualization is all about. I did that! I created that beautiful thing. I used my problem-solving skills to keep the pests at bay naturally. Growing this food has brought great joy to my life and boosted my self-esteem and confidence. This broccoli really is a huge achievement. This is the upper part of the needs hierarchy that you cannot learn from a book or a DVD lecture. You just have to do it, and the accomplishment will dwarf anything else out there.

I have been doing workshops on the coming challenging times for many years. What I've written about in these pages I condense into about three hours in a workshop. Many of the people who attend have been watching things happen in the world and have started to wonder about how it all fits together. They see a news report on environmental problems. They read about the challenges of feeding 6 ½ billion people. There's a fire at a local refinery and they arrive at the gas station with an empty gas tank only to find there is no gas. On one level they've been aware of what's going on, but they haven't wanted to think about where the big picture takes them. My workshop forces them to do it. As in the movie *The Matrix* they come and take the red pill and see the sorts of challenges we really face.

When I finish a workshop and try and encourage some discussion I usually see a lot of dazed and confused faces. I think people are going through a period of grief. They suddenly realize that things have changed and that they're probably not going to go back to the way they were. I think there's a bit of denial, but since they've attended the workshop they were already partway to this point. I think some are angry. How did governments let this happen? Why didn't we all listen to the Club of Rome in the 1970s and determine what the true carrying capacity of the planet is and try and keep our population there? I have no doubt that they are depressed. But as I draw them out I can usually start seeing a glimmer of hope and acceptance. Okay, things may seem bad but only if you define yourself within the current economic paradigm. We can start to redefine our lives and what has meaning and we may be okay. I see them not only start to accept this new reality but I sense that many are actually energized to change. Some have had a long-range plan for years but never articulated it. They never wanted to set a time for it, but now they're motivated. Now they have to get going.

One of the workshop attendants worked as a purchaser of fresh fruits and vegetables for a large grocery chain. He came to the workshop as the price of oil was passing $100 a barrel. When confronted with the reality of peak oil he realized that his job was premised on getting 20 long-haul trucks a week from the southern United States to the north so that the store could have the same fresh fruits and vegetables 52 weeks a year. I think he had an epiphany that long-term his job would not be secure. He didn't seem freaked out; he seemed excited that he could start refocusing his career to deal with the new reality.

I've had many people at workshops who own rural property that they one day hope to retire to. Suddenly waiting until retirement doesn't seem so important. Making that place habitable and independent is their goal in life and this information has spurred them into action. I had a group of four women in my workshop who had been talking about the various challenges the world faces for many years and asked if they could come up to our place to learn more. They came for a weekend to find out about living off the grid. Their goal was to build an intentional community where each would have her own private space but where they would have a common shared space as well. They were using the model of building a community where each would contribute her unique skills. Two of the women were engineers. That's a pretty good place to start from. They came as oil was approaching $150 a barrel and the U.S. housing market was beginning to collapse. They wanted a date! When was it all going to come unglued? "Give me a date so I can prepare."

I couldn't give them a date. Humans are very resilient and we have many institutions that should be able to help us weather these challenging times. But peak oil sets a limit on how good the times will be that we may return to. It says the future will be different, so get prepared.

I think one of the main things people get from my workshops is just how happy I am. "What's the point of going on if things are such a mess?" is a common question. All I can say is that humans have lived under far worse conditions than we are experiencing and many of them were happy while they were at it. I don't believe in the Buddhist mantra that all life is suffering. Sure it can be hard, but that doesn't mean you can't be happy. Have you seen pictures of kids from other countries? Less fortunate kids, sometimes living in squalor. They so often seem happy. Their happiness seems legitimate. It's not put on. Then you read about how many of our kids from developed nations are on anti-depressants. Aren't their personal TVs, video games, iPods, cell phones, endless entertainment, limitless food, and infinite choice making them "happy"? Apparently not.

The more I analyze it the more I honestly believe my happiness is inversely related to my income. Driving across the country for two months on our honeymoon 25 years ago and living on $1,000 was without question one of the best times in my life. Over the last few years as we've eliminated all our traditional sources of income and started publishing books and DVDs on sustainability full-time our income has been on a steady downward trajectory. My income last year was appalling. I would be able to make more money at a fast-food restaurant. And yet I've never been happier. I'm so grateful that I worked my ass off for many years earning a better income and paying off my mortgage so I could come to this place and live with so little.

My retirement fund is almost gone and I'm only 50. Each year I take out a bit more. Two years ago it was for the new wind turbine. Last year it was for the solar domestic hot water heater. But each one of those hard assets made my house more independent so that as I grow older I'll have to spend less and less money. As Dan Buettner points out in his book *The Blue Zone,* in cultures where people routinely live the longest they do not have a word for retirement[3]. They do not retire. They just keep on working, growing food, and being active for as long as they live. And they live much longer than we do. Apparently retirement, which sounds good while you're slogging it out in some cubical in an office tower, sounds great in principal, but when you get there it's a huge disappointment. You no longer have meaning in your life. Your mind and your body don't need to be as active and they start to wind down. Seniors homes start looking like human warehouses.

So if like so many of us you've seen the stock market crash and economic downturn devastate your retirement fund, look at the bright side. It probably means you'll be living longer. Why? Because you're going to have to keep active. You're going to have to keep growing your own food and coming up with ways to make your home more resilient to energy shocks. Your retirement is going to be one of new pursuits, activity, challenges, and great joy in your ongoing accomplishments. While I feel blessed to have had some major accomplishments in my life—raising two wonderful daughters, running my own business for 25 years, living off the grid, publishing books that help people live sustainably—my greatest accomplishments still occur on a daily basis. The load of manure I just spread on the garden. The rotten hay bales I just scrounged which will break down and help condition my soil. The pile of wood I cut and split and stacked today. I can stand back and look at that wood and just beam. I did that! That's almost two weeks of next winter's heat, keeping my house toasty warm. Sometimes I'll go outside after dinner just to look at that pile. It's the real deal. It's real work. It's heat and security. It's independence. It was created through the wonder of trees absorbing carbon dioxide and through my burning of calories to cut it and split it and chop it and pile it. I do not believe you can get that same feeling from pushing pixels around a computer screen or whacking a little white ball into a hole.

I believe as you begin your journey towards independence you'll understand what I mean. I believe you will change your definition of happiness and it won't be whether or not you made it to Costa Rica this year or how many shoes are in your closet. It will be how many buckets of potatoes you got into the root cellar this fall and how low your natural gas bill is now that you've installed a solar panel on your roof to heat your water. Your grandparents on the farm did not have much "stuff"; they had very little material wealth, but their lives were much richer than ours and I believe they experienced on a daily basis the joy that comes from independence. And with every step you take towards your own independence they will be gazing down upon you beaming!

22
Thriving During Challenging Times

Happiness comes when we test our skills towards some meaningful purpose.
John Stossel

We must have courage to bet on our ideas, to take the calculated risk, and to act. Everyday living requires courage if life is to be effective and bring happiness.
Maxwell Maltz

When one door of happiness closes, another opens; but often we look so long at the closed door that we do not see the one which has opened for us.
Helen Keller

Michelle and I have experienced some of the dislocation that many people will encounter in the future or perhaps have already experienced with the economic downturn. Living four miles from the nearest electricity line as we do also means we have no phone line coming to our house. When we first moved off the grid we had no cellphone service, so we used a high-tech piece of equipment called an Optaphone. This was a point-to-point phone line extender. We installed a utility pole at the end of the phone line and attached a radio device that converted the phone line to radio waves. We had a receiver in our house to convert the radio signal back to a phone signal.

This system was problematic but worked most of the time. It was an extremely slow Internet connection and tied up our phone line when we were online. So we eventually installed a satellite Internet connection. This was much faster and we were gradually able to convince most of our customers that the most reliable way to communicate with us was via email.

As luck would have it, as our Optaphone was coming to the end of its natural life we had a large thunderstorm roll through the area. When it was over, several of our computers were damaged, as was our satellite Internet system. Lightning had struck the ground and an underground Ethernet cable between our house and our office took the electrical surge into the Ethernet ports in the computers. We had hoped to take a few months to research an alternative to our phone system while relying on an older bag-style cellphone. But now our Internet was also not working properly, so we were without reliable phone and Internet communication to the outside world. Trying to run a business under these circumstances was very difficult.

To add to the stress our eldest daughter Nicole was at the far end of the province of Quebec on a French exchange program for the summer. Nicole is a wonderfully adventuresome daughter, and even though she had almost no French experience she jumped fearlessly headfirst into the program anyway. But when she got there the magnitude of the challenge suddenly dawned on her and she discovered the program did not allow her to speak any English, so it was sink or swim. So there she was, ten hours from home, unable to speak French and stuck for five weeks with people who were not cutting her any slack.

I was spending hours on our expensive bag phone speaking to the highest level of tech support at our satellite Internet provider and trying to figure out a new phone system while trying to help our daughter through her exchange challenges and keep providing work to customers from whom we earned our income. It was an insanely stressful and disorienting time. When you have come to rely on and take for granted certain things like a phone and the Internet, it completely discombobulates you when they stop working. Up is down and left is right and it's not easy to stay centered. It's sometimes difficult to stay focused on the goal. It's hard to break it down into smaller chunks of manageable tasks and not get overwhelmed by the tsunami of problems rolling over you.

But you have to.

This is how much of what will start happening in the future will affect us. We will not have control over the price of gas or whether or not we can actually get it for our cars. We won't be able to control the price of the natural gas that heats our homes or whether the grocery store has enough food for us. There will be shortages. These will be controlled externally. All we can do is anticipate these scenarios and try and create systems to deal with them. Like living close enough to work that you can walk. Or running your own business out of your home. Or heating your home with a geothermal heat pump that doesn't require fossil fuels. Or growing so much food in your garden that you can live off its bounty for the winter. That's the way you deal with the stress. Know that it's coming and prepare yourself. Understand that it will be disorienting but that with the right attitude you can handle it as it comes.

You have to start breaking it down into manageable chunks. You should start with taking a big-picture inventory of where you're at and where you want to go. Then you have to decide how you're going to get there. Where will you be living? What kind of house? How much debt do you have? How quickly can you get out of debt? Then you can start doing the fun and easy stuff. Making your house more energy efficient to save money. Starting your garden. Scrounging leaves from the neighbor's curb for your compost. Grabbing that glass you saw three blocks over for your homemade greenhouse. Saving to get that solar domestic hot water unit installed.

As you start making progress towards your goals you may want to start creating a bit of a media blackout. If you keep watching and listening to the news and it all seems pretty bleak it can be discouraging. The media will not necessarily be giving you the real information you need anyway since there's very little upside to the story. There's no way for them to put a spin on it to make money. So try and ignore it. Use all those clichés—"Turn your mainsail to the wind," "Onward and upward," "Put your head down and plow forward"—to get you moving and keep you moving.

I hope this book has set out some strategies to help you come up with a plan for your personal independence. Having a plan sounds like a good idea but lots of us don't have one. We think we do but we have trouble articulating it. Sometimes putting your plans in writing will help. Put your five-year plan down on paper and keep referring to it. Keep updating as you go. Be sure to leave the things you accomplish on your list as you do them, like paying off all your credit card debt, so you can see what great progress you've made.

There are lots of examples of people who had a plan for surviving very challenging situations. I've read about people who survived horrific plane crashes. Often they say, "I sat down at my seat and checked out the emergency exits. I found the closest exit. I said to myself, "If this plane goes down I'm getting to that door and I'm getting out that door. Nothing will stop me." And when there was chaos and smoke they got to that door and they got out. Many people who survived the WTC attacks knew where the staircases were because they did lots of fire drills. How uncool is it to actually follow through with a fire drill, but as it turns out it saved lives. Some people knew the staircases in those towers so well that when they got down from the floors above where the planes had hit and the staircase was blocked they went back up and came down another set. They had made a plan and they stuck to it and nothing stopped them from following through.

Make a plan. Practice your plan. Rehearse your plan. Stick to your plan. You can be flexible with your plan. Things will change, but once you set that goal stick to it. Whether it's getting out of that burning plane or getting out of debt and buying a hobby farm in the next five years, set your sights and stay on course.

Life is a big adventure. It takes you in many different directions. It has ups and it has downs and sometimes you have to ask yourself how you got to where you're at. I hope you are at a good point in your life and things are going well for you. I think the current economic challenges have left many of us in places we'd rather not be. I hope I have provided some guidance for where you might want to go. Key to everything will be getting out of debt and staying there. I know, it's going to be a rough road to travel, but you've got to do it. Everything else depends on it. Being debt free is the foundation on which independence is built.

You do not need to be wealthy or have a high income to become financially independent.

You need to be frugal and have a plan to avoid debt. It's that simple.

Michelle and I had a plan to get out of debt, including our mortgage, and we totally committed to it. We didn't buy anything we didn't absolutely need. This started us on a path of disciplined saving. Then we set our goal on finding a rural property to move to. When we found it we did everything we could to make sure we could transition our home-based business to our new home three hours away. It was a huge amount of effort but it worked. While we were working brutal hours to keep an income, we were upgrading our solar panels, replacing and upgrading many systems in the house, expanding gardens, installing a new wind turbine, and doing all the time-consuming things that come with 150 acres, including cutting our own firewood and reducing how much gas we used to haul it back to the house. It's a crazy amount of work, and we could probably make more money if we just worked harder at earning an income and purchased firewood from someone else.

But heating with wood cut sustainably from our property is the embodiment of all our beliefs right now. It is a carbon-neutral way to heat. When we cut dead trees the smaller ones grow back quickly to take their place. I love splitting and stacking firewood. Sometimes the best part of my life is staring at firewood I've just cut. There is no sense of accomplishment like it. I usually have next winter's firewood by May. Staring at that pile says I've got my home heated next winter. It says I don't have to worry about war in the Persian Gulf restricting oil exports and there not being fuel oil for my furnace. And I don't have to worry about a super spike in natural gas taking most of my excess income next winter because North America is finding less natural gas.

Next winter my lights will be on, my fridge will be keeping food cool, my freezer will be keeping vegetables from my garden frozen, and my computers and Internet will keep me linked to the outside world 24 hours a day, 7 days a week without the possibility of any interruption, because all the power is generated by my solar panels and wind turbine. Next winter I'll be having showers and baths with water heated by the sun and my wood stove, regardless of whether the propane truck decides to come all the way down my road to fill up my tanks. I get to stay clean without being at someone else's mercy.

There really is something to be said for this sense of independence. It's very relaxing. It's very liberating. Our income over the last ten years has been falling steadily by choice. It went down when we started giving up customers three hours away whose business models didn't fit our definition of sustainability. It came down dramatically as we shifted to publishing only books on sustainability. It continues to fall to a point where we're getting pretty close to living below the poverty line. But the poverty line is based on people having to pay for accommodation, be it rent or a mortgage, and purchase electricity and gas for heat, food for groceries, and all the other things most North Americans have to purchase.

Accommodation? Some property taxes and insurance we can cover. Electricity? Make our own. Heat? Cut our own. Food? Grow our own. Transportation? Burn some gas but use the electric bike and stay home more and more. Entertainment? Mergansers in the pond. Used books from the secondhand store when we splurge! It's a good life. We live like kings, but society says we're poor. And so it takes pity on us. And we pay very little income tax. In fact when the price of heating oil and natural gas spike, the government often sends us a check because we make so little money and they assume we need it to stay warm. We spend it locally, but not on heat. Maybe some hot food at the nearest Indian restaurant.

There is no way to describe strolling over to the garden and seeing your own food. Or looking at your firewood and knowing you'll be warm next winter. Or seeing your solar panels and knowing the lights and the fridge will never go out. I tell you this not to brag, but hopefully to motivate. I emphasize how little money we make not to win your sympathy but to show you that if you work towards independence the payback is living like a king on very little money with way less stress. There will be no vacations to warm beaches in the winter for us. And that's okay. There will never be that mid-life-crisis sports car, because there is no money for it. No problem, the electric bike which maxes out at 25 miles per hour is perfect! Plus the sun recharges it when we get home.

You now have the knowledge and I hope the motivation to start making yourself be more independent. You have the financial resources to do so if you change your priorities. A $10,000 slate countertop will not help during challenging times. Investing in solar panels will. Once you make your plan, stick to it. If you start on the path there is no downside. If you follow through with this you'll have created a more sustainable lifestyle where you've reduced your footprint on the planet and started to take joy in everyday living rather than planning for some mythical retirement that may never happen or that turns out to be a disappointment

There is a wonderful band called The Wailin' Jennys. A great name and three exceptionally talented women. Band member Ruth Moody wrote a wonderful song called "Heaven When We're Home"[1]:

It's a long and rugged road
and we don't know where it's headed
But we know it's going to get us where we're going
And when we find what we're looking for
we'll drop these bags and search no more
'Cuz it's going to feel like heaven when we're home.

We worked hard to find this place.

We worked hard to pay it off.

We have dropped our bags and we're not leaving and it feels like living in heaven on earth.

I wish you all the best on your journey. I hope you can find your direction and set your goals on a target and achieve them. I think becoming radically independent will serve you very well in the challenging times ahead. It can be a long and rugged road, but it's worth the effort. When you get that independence remember that the spirit of your grandparents and great grandparents will be there beside you, very proud to see you back in the "living within your means" fold.

So what are you waiting for? Sit down at your computer. Make a budget. Look at your debt. Set a timeline on when you want to be debt free. Decide where you want to be living in five years. Pick up a seed catalog and order seeds for your vegetable garden. Get some prices on solar panels. Take a long walk to think all this through. Ride your bike to work on Monday. Have a cup of coffee with a friend or your spouse. Wonder and be amazed at the distance those coffee beans have traveled. Marvel that you can afford such a luxury.

Life is indeed good.

Notes

Chapter 3   Economic Collapse

1 Willem Buiter, "Buiter: Welcome to a World of Diminished Expectations," *FT.com*, 5 August 2008 <http://www.ft.com/cms/s/0/0bbf7a38-631d-11dd-9fd0-0000779fd2ac. html> (8 July 2009).

2 Richard W. Fisher, "Storms on the Horizon," *Federal Reserve Bank of Dallas*, 28 May 2008 <http://www.dallasfed.org/news/speeches/fisher/2008/fs080528.cfm> (8 July 2009).

Chapter 4   Peak Oil

1 Matthew R. Simmons, *Twilight in the Desert: The Coming Saudi Oil Shock and the World Economy* (New Jersey: John Wiley & Sons, 2005).

2 *International Energy Agency,* "World Energy Outlook 2008 Edition," 2008, <http:// www.iea.org/Textbase/publications/free_new_Desc.asp?PUBS_ID=2056> (8 July 2009).

3 George Monbiot, "At Last, A Date," *Monbiot.com*, 15 December 2008 <http://www. monbiot.com/archives/2008/12/15/at-last-a-date/> (8 July 2009).

4 Robert L. Hirsch, "Peaking of World Oil Production: Impacts, Mitigation & Risk Management," *The National Energy Technology Laboratory*, February 2005 <http://www.netl.doe.gov/publications/others/pdf/Oil_Peaking_NETL.pdf> (8 July 2009).

5 Ibid.

Chapter 5   Peak Food

1 *My Own Bag*, "Bags by the Numbers," 2005 <http://www.myownbag.com/activism. html> (8 July 2009).

2 Gretel H. Schueller, "Eating Locally," *Discover Magazine*, 1 May 2001 <http://discovermagazine.com/2001/may/feateatlocal> (8 July 2009).

3 Karen Barlow, "Food riots 'an apocalyptic warning'," *ABC News*, 14 April 2008 <http://www.abc.net.au/news/stories/2008/04/14/2215873.htm?section=world> (8 July 2009).

4 Lester R. Brown, "World Facing Huge New Challenge on Food Front," *Earth Policy Institute*, 16 April 2008 <http://www.earthpolicy.org/Updates/2008/Update72.htm> (8 July 2009).

5 Keith Bradsher, Andrew Martin, "Hoarding Nations Drive Food Costs Ever Higher," *The New York Times,* June 30, 2008 <http://www.nytimes.com/2008/06/30/business/worldbusiness/30trade.html? pagewanted=2&_r=4&hp&adxnnlx=1214923043-KafOaaHWqfaeO1VaOfD%205Q> (8 July 2009)

6 Rosie Boycott, "Nine meals from anarchy - how Britain is facing a very real food crisis," *Daily Mail*, 7 June 2008 <http://www.dailymail.co.uk/news/article-1024833/Nine-meals-anarchy--Britain-facing-real-food-crisis.html> (8 July 2009).

Chapter 6   Peak Water

1 *International Energy Agency*, "World Energy Outlook 2005 Edition- Middle East and North Africa Insights," 2005 < http://www.iea.org/Textbase/publications/free_new_Desc.asp?PUBS_ID=1540> (13 July 2009).

2 Jim Tankersley, "California farms, vineyards in peril from warming, U.S. energy secretary warns," *Los Angeles Times*, 4 February 2009, <http://articles.latimes.com/2009/feb/04/local/me-warming4> (8 July 2009).

3 Jeff Fleischer, "Blue Gold: An Interview with Maude Barlow," *Mother Jones*, 14 January 2005 <http://www.motherjones.com/news/qa/2005/01/maude_barlow.html> (8 July 2009).

4 Fred Pearce, "Water Scarcity: The Real Food Crisis," *AlterNet*, 9 June 2008 <http://www.alternet.org/water/87234/> (8 July 2009).

5 Daniel Zimmer and Daniel Renault, "Virtual Water in Food Production and Global Trade," *Food and Agriculture Organizations of the United Nations*, 2003 <www.fao.org/nr/water/docs/VirtualWater_article_DZDR.pdf> (8 July 2009).

Chapter 7   Climate Change

1 *International Energy Agency*, "World Energy Outlook 2008 Edition," Executive Summary p.37, 2008, <http://www.iea.org/Textbase/publications/free_new_Desc.asp?PUBS_ID=2056> (8 July 2009).

2 *National Snow and Ice Data Center*, "Arctic Sea Ice Shatters All Previous Record Lows," 1 October 2007 <http://nsidc.org/news/press/2007_seaiceminimum/20071001_press-release.html> (8 July 2009).

3 Gwynne Dyer, *Climate Wars* (Canada: Random House, 2008).

Chapter 8   Why It's Different This Time

1 *The Age*, "Golden Gate to nowhere," 19 December 2008 <http://business.theage.com.au/business/world-business/golden-gate-to-nowhere-20081219-71u1.html?ref=patrick.net> (8 July 2009).

2 Tom Abate, "Poor families' 'safety net' is wearing thinner," *San Francisco Chronicle*, 24 December 2008 <http://www.sfgate.com/cgi-bin/article.cgi?f=/c/a/2008/12/23/BU6114U6UO.DTL> (8 July 2009).

3 Kristin Kloberdanz, "The Great California Fiscal Earthquake," *TIME*, 8 January 2008 <http://www.time.com/time/nation/article/0,8599,1870299,00.html?ref=patrick.net> (8 July 2009).

4 Dmitry Orlov, "Post-Soviet Lessons for a Post-American Century," *The Wilderness Publications*, 2005 <http://www.fromthewilderness.com>

Chapter 9   Where to Live

1 *OECD*, "Growing Unequal? : Income Distribution and Poverty in OECD Countries," October 2008 <www.oecd.org/els/social/inequality> (8 July 2009).

2 Matthew Benjamin, "Americans See Widening Rich-Poor Income Gap as Cause for Alarm," *Bloomberg.com*, 13 December 2006 <http://www.bloomberg.com/apps/news?pid=20601087&sid=atGy4g3gcN4I&refer=home> (8 July 2009).

3 Garth Turner, *After the Crash* (Key Porter Books, 2009).

Chapter 11   Heating

1   Jason Leopold, "5 Years After Blackout, Power Grid Still in 'Dire Straits'," *Atlantic Free Press*, 17 August 2008 <http://www.atlanticfreepress.com/content/view/4786/81/> (13 July 2009).

Chapter 13   Food

1   *Colorado State University Extension*, "Food Safety and Storage for Emergency Preparedness," 2009 <http://www.ext.colostate.edu/Pubs/emergency/fdsf.html> (8 July 2009).

2   *International Year of the Potato*, "Why potato?" 2008 <http://www.potato2008.org/en/aboutiyp/index.html> (8 July 2009).

Chapter 16   Transportation

1   *TVO*, "The End of Cheap Oil," The Agenda, 17 March 2009 <http://www.tvo.org/cfmx/tvoorg/theagenda/index.cfm?page_id=401&action=viewthread&forum_thread_id=7873&forum_id=43> (9 July 2009).

2   *Bureau of Transportation Statistics*, "Table 1-11: Number of U.S. Aircraft, Vehicles, Vessels, and Other Conveyances," n.d. <http://www.bts.gov/publications/national_transportation_statistics/html/table_01_11.html> (9 July 2009).

Chapter 17   Health Care

1   M. Thorogood, J. Mann, P. Appleby, and K. McPherson, "Risk of Death from Cancer and Ischaemic Heart Disease in Meat and Non-Meat Eaters," *British Medical Journal* 308 (June, 1994): 1667-70.
    J. Chang-Claude, R. Frentzel-Beyme, and U. Eilber U, "Mortality Patterns of German Vegetarians after 11 Years of Follow-Up," *Epidemiology* 3 (1992): 395-401.
    J. Chang-Claude and R. Frentzel-Beyme, "Dietary and Lifestyle Determinants of Mortality among German Vegetarians," *International Journal of Epidemiology* (1993): 228-36.

2   N.D. Barnard, A. Nicholson, and J.L.Howard, "The Medical Costs Attributable to Meat Consumption," *Preventive Medicine* 24 (1995): 646-55.

3   World Cancer Research Fund, *Food, Nutrition, and the Prevention of Cancer: A Global Perspective* (Washington, DC: American Institute for Cancer Research, 1997).

4   R.J. Barnard, L. Lattimore, R.G. Holly, S. Cherny, and N. Pritikin, "Response of Non-Insulindependent Diabetic Patients to an Intensive Program of Diet and Exercise," *Diabetes Care* 5 (1982): 370-4.

Chapter 18   Safety and Security

1   Don Lehman, "Gun Sales Going Ballistic," *Glens Falls (NY) Post-Star*, 20 March 2009 <http://www.poststar.com/articles/2009/03/20/news/local/14551013.txt> (9 July 2009).

2   Henry Pierson Curtis, "Gun Dealers Experiencing Shortage of Bullets," *Orlando Sentinel*, 10 February 2009 <http://www.orlandosentinel.com/news/local/orl-bullets1009feb10,0,2201778.story> (9 July 2009).

3 Peggy Noonan, "There's No Pill for This Kind of Depression," *Wall Street Journal,* 13 March 2009 <http://online.wsj.com/article/SB123689292159011723.html?ref=patrick. net> (9 July 2009).

4 James Howard Kunstler, *The Long Emergency* (New York: Grove Press, 2005).

5 Eli Lake and Stephen Dinan, "US intel: Economy #1 security threat," *The Washington Times,* 12 Feb 2009, <http://www.washingtontimes.com/news/2009/feb/12/us-intel-economy-greatest-security-threat/> (9 July 2009).

6 *The Wall Street Journal,* "Tilting at Windmill Jobs", 3 July 2009 <http://online.wsj.com/article/SB124657739768489217.html#mod=loomia?loomia_ si=t0:a16:g2:r1:c0.346904:b26193184> (9 July 2009).

7 Nathan P. Freier, "Known Unknowns: Unconventional 'Strategic Shocks' In Defense Strategy Development," *Strategic Studies Institute (SSI),* November 2008, p.32 <http://www.strategicstudiesinstitute.army.mil/pubs/display.cfm?pubID=890> (9 July 2009).

Chapter 19   Money

1 *The 11th Hour: Turn Mankind's Darkest Hour Into Its Finest,* narrated by Leonardo DiCaprio (Warnervideo, 2007).

Chapter 20   Mediums of Exchange

1 William Bonner, *Empire of Debt* (New Jersey: John Wiley & Sons, 2006), 328.

2 Richard Fisher, "Storms on the Horizon," *Federal Reserve Bank of Dallas,* 28 May 2008 <http://www.dallasfed.org/news/speeches/fisher/2008/fs080528.cfm> (9 July 2009).

3 Ron Paul, "What the Price of Gold Is Telling Us," *LewRockwell.com,* 27 April 2006 <http://www.lewrockwell.com/paul/paul319.html> (9 July 2009).

4 Kevin Phillips, "Why the Economy is Worse Than We Know," *Harper's Magazine,* 1 May 2008 <http://www.mindfully.org/Reform/2008/Pollyanna-Creep-Economy1may08.htm> (9 July 2009).

5 Brook Larmer, "The Real Price of Gold," National Geographic, p.43, January, 2009.

6 *United States Mint Online Catalog,* "American Eagle Gold Uncirculated Coins," 2009 <http://catalog.usmint.gov/webapp/wcs/stores/servlet/CategoryDisplay?langId=-1&storeId=10001&catalogId=10001&identifier=1150> (9 July 2009).

Chapter 21   Living Happily in the New World

1 Reprinted by arrangement with Penguin Books, a member of Penguin Group (USA) Inc., from *Your Money or Your Life* by Vicki Robin & Joe Dominguez with Monique Tilford. Copyright © 2009 by Your Money or Your Life by Vicki Robin and Monique Tilford.

2 Gregg Easterbrook, *Progress Paradox: How Life Gets Better While People Feel Worse* (New York: Random House, 2003).

3 Dan Buettner, *The Blue Zone* (Washington, D.C.: National Geographic, 2008).

Chapter 22   Thriving during Challenging Times

1 The Wailin' Jennys, "Heaven When We're Home," *40 Days,* Red House, 2004.

Index

A

Adjustable Rate Mortgages 17
airline industry 84
air travel 85, 202
auto industry 84

B

backup generator 72, 281
Barack Obama (also see Obama) 20, 42, 78, 156, 223, 224, 225, 263
Barlow, Maude 42, 307
barter 10, 74, 75, 91, 93, 266, 267, 268, 269
batteries 120, 123, 125, 126, 128, 200, 203
Bernacke, Ben 17
bicycle 67, 201, 202
biodiesel 82, 91, 98, 200
Biodiesel Basics and Beyond 98, 200, 315
blackout 98, 99, 108, 118, 186, 281, 282, 301
Bonner, William 263, 309, 314
Brown, Lester 37, 43

C

CO$_2$ (see also carbon dioxie) 23, 99, 112, 202
California 9, 37, 42, 53, 54, 55, 70, 97, 119, 143, 220, 284, 306, 307
Campbell, Colin 29
carbon dioxide (see also CO$_2$) 23, 47, 49, 101, 131, 137, 181, 297
cash 12, 14, 91, 224, 227, 233, 235, 237, 241, 246, 247, 249, 251, 252, 260, 270, 272, 277, 278, 279, 280, 281, 282
catalytic wood stove 101, 102
ceiling fans 112
China 15, 28, 37, 38, 39, 43, 45, 85, 194, 244, 245
Chu, Steven 42
climate change 37, 54
clothesline 110, 111, 116
Club of Rome 33, 43, 295
coal 23, 59, 62, 63, 68, 99, 107, 110, 131, 170, 198, 212
cold frame 168
Collateralized Debt Obligations 9, 17
community 75, 76, 81, 93, 121, 156, 158, 197, 205, 210, 218, 226, 227, 231, 261, 292, 295

Community Supported Agriculture (see also CSA) 88, 149, 158
compact fluorescent light bulb 109, 110, 147
compost 157, 159, 160, 161, 162, 163, 164, 165, 167, 168, 174, 176, 177, 178, 181, 301
composting toilets 187
credit card 12, 14, 245, 246, 247, 248, 249, 250, 252, 253, 255, 263, 277, 289, 301
credit card debt 245
CSA (see also Community Shared Agriculture) 149, 150, 153, 158
Cuba 57, 58, 89
cutworm 172, 174
cycling 6, 41, 93, 188, 189, 196, 206, 213

D

Dallas Federal Reserve Bank 21, 264
Darley, Julian 95, 314
debt 11, 12, 13, 14, 18, 19, 20, 21, 23, 52, 80, 81, 82, 86, 209, 236, 237, 239, 245, 246, 247, 248, 249, 252, 254, 260, 262, 263, 264, 274, 282, 283, 287, 289, 301, 302, 305
debt-to-GDP ratio 13
Deffeyes, Kenneth 29, 30, 314
deflation 75, 77, 78, 79, 80, 81
Dent, Harry 261
depression 19, 51, 52, 53, 81, 234, 261, 264, 273, 286
diesel 24, 36, 40, 57, 67, 68, 84, 91, 92, 97, 122, 143, 150, 155, 177, 194, 199, 200, 201, 205
Dominguez, Joe 287, 309
Douglas, Tommy 207, 208
drinking water 42, 44, 183, 184, 185, 188, 220
drip irrigation 177
Dyer, Gwynne 49, 307

E

Easterbrook, Gregg 292, 309
efficiency 77, 82, 101, 108, 110, 112, 113, 114, 118, 131, 199, 202, 216, 252, 253
electric bike 203, 204, 206, 303, 304
electric chain saw 103
electric vehicles 194, 198

Emergency Banking Act 270
Emergency Fund 253, 254
EnerGuide 113
energy audit 114, 252
EnerStar 113
EPA-certified 73, 101, 105
EpiPen 221, 222
Exchange Traded Funds (ETFs) 271, 272

F

Fed (see also Federal Reserve) 15, 16, 17, 54, 78, 265
Federal Deposit Insurance Company 260
Federal Reserve 15, 19, 21, 76, 78, 234, 242, 264, 270, 306, 309
fertilizer 33, 36, 66, 71, 137, 163, 164
fiat currency 11, 12, 263
firewood 6, 67, 73, 74, 76, 102, 103, 104, 105, 216, 267, 268, 286, 303, 304
Fisher, Richard W. 21, 264, 306
Five-Year Plan 6, 62, 65, 238, 240, 241, 245
Food and Agricultural Organization of the United Nations 41
food pyramid 133, 135
food shock 138
Fortune Magazine 41, 133
freeze-dried foods 152
freezer 120, 125, 138, 144, 145, 167, 180, 266, 303

G

garden plot 157, 158
General Motors 9, 82, 83
generator 5, 69, 72, 120, 121, 122, 123, 125, 128, 186, 281
geothermal 100, 101, 301
gold 10, 11, 12, 14, 15, 16, 20, 161, 191, 233, 263, 266, 269, 270, 271, 273, 274, 275, 276, 277, 278, 279, 280, 281
gold standard 11, 12, 15, 263, 269
graywater 190
Great Depression 59, 79, 259, 261, 264, 267
Green Energy Plan 206
greenhouse gas emissions 47, 155
green jobs 89
Greenspan, Alan 15, 16, 17, 18, 19, 76, 77, 234, 241, 242, 272
Gross Domestic Product (GDP) 12, 13, 53, 264

gun 223, 224, 228, 229, 230, 231, 252

H

handgun 228, 230
Hard Assets 266
hard landing 55, 152, 156, 224, 267
hay 71, 93, 137, 160, 161, 162, 164, 175, 176, 205, 222, 297
health care 14, 51, 58, 184, 207, 208, 209, 210, 212, 213, 217, 219, 222
heat pump 73, 92, 100, 104, 294, 301
hierarchy of needs 60, 95, 293
home alarm 227
home heating oil 23, 97, 99
home ownership 76
Home Power magazine 89, 90
Homer-Dixon, Thomas 60, 314, 315
horses 68, 71, 82, 204, 205, 206
hot water 7, 64, 92, 98, 100, 107, 111, 113, 114, 115, 116, 117, 118, 129, 139, 244, 252, 254, 255, 260, 261, 292, 294, 296, 301
Hubbert, Marion King 24, 25, 30, 314
hurricane 47, 48, 49, 119, 126, 141, 185, 230, 261, 281
hybrids 84, 176, 194, 199, 202

I

IEA 28, 30, 31, 306
inflation 75, 77, 78, 79, 81, 119, 131, 132, 265, 266, 270, 272, 281
intentional communities 75, 227
Intergovernmental Panel on Climate Change 47
International Energy Agency 28, 41, 48, 306, 307 (see also IEA)
inverter 117, 127, 128, 186, 204, 322
investing 98, 105, 107, 118, 128, 131, 152, 188, 256, 259, 286

J

jerry cans 122, 160, 266

K

Katrina 47, 49, 225
Kemp, William H. 98, 108, 116, 117, 121, 129, 130, 200, 253, 315, 322
Kunstler, James Howard 225, 308, 314

L

Limits to Growth 33, 34
line of credit 11, 19, 246, 249, 250, 251, 252
Liquid Natural Gas (LNG) 30, 96, 97

M

Maldives 47, 49, 202
Malthus, Thomas 33, 38
manure 71, 163, 164, 165, 170, 205, 297, 322
Maslow 60, 95, 293
Matlack, Larry 35
Medicare 20, 87, 264
Monbiot, George 29, 202, 306, 315
money supply 15, 18, 78, 79, 265
monthly budget 69, 155, 254, 255
mortgage 9, 11, 12, 14, 16, 17, 18, 54, 59, 74, 76, 77, 80, 81, 82, 94, 209, 236, 237, 238, 239, 240, 241, 242, 243, 245, 246, 249, 250, 251, 252, 254, 255, 257, 262, 282, 289, 296, 302, 303
mortgage payments 9, 59, 76, 237, 239
Mother Jones 42, 307
motorcycles 193, 200
mountain bike 206
mutual funds 271

N

"non-essential" loads 120
National Foundation for Credit Counseling 247, 248
National Snow and Ice Data Center 48, 307
natural gas 23, 29, 30, 33, 36, 54, 66, 68, 74, 77, 82, 88, 95, 96, 97, 99, 104, 105, 107, 111, 114, 116, 117, 121, 122, 129, 137, 144, 155, 198, 209, 252, 253, 254, 268, 297, 301, 303, 304
NINJA 16, 272
Nixon, Richard 12, 13, 270
noncatalytic wood stove 101

O

Obama 20, 42, 78, 156, 223, 224, 225, 263
off-grid 6, 63, 82, 125, 129, 187
Ogallala Aquifer 44
OPEC 26, 27, 193
Organisation for Economic Cooperation and Development (OECD) 28, 69, 306, 307

Orlov, Dimitry 56, 307

P

Paiken, Steve 194
pantry 35, 40, 121, 139, 140, 141, 142, 144, 152, 153, 154, 269
paper dollars 152, 266, 268, 273, 274, 276, 280
payback 2, 109, 115, 116, 117, 118, 130, 131, 238, 253, 304
paying off debt 236
peak oil 24, 25, 26, 28, 29, 30, 31, 36, 54, 55, 57, 64, 68, 82, 83, 91, 95, 97, 116, 129, 144, 155, 194, 195, 197, 209, 216, 225, 295, 296
Pearce, Fred 43, 44, 307
pellet stoves 104
pension funds 9
pests (garden) 139, 142, 147, 171, 172, 178, 294
phantom loads 113, 114
photovoltaic 117, 153
Phillips, Kevin 265, 309, 314
Physicians Committee for Responsible Medicine (PCRM) 211
Pickens, T. Boone 29
Plan B 31, 55, 58, 68, 73, 104, 105, 171
plant-based diet 135, 137, 138, 153, 213, 214, 217
population 13, 33, 34, 36, 39, 40, 41, 43, 51, 54, 56, 57, 66, 67, 77, 135, 153, 195, 196, 208, 222, 268, 295
potatoes 66, 121, 135, 137, 141, 147, 148, 149, 151, 153, 169, 172, 173, 179, 181, 210, 215, 219, 246, 268, 297
precious metals 10, 263, 270, 275, 276, 277
President Obama 224, 263
pressure canner 144
price of oil 30, 65, 84, 85, 88, 287, 295
propane 24, 64, 92, 117, 122, 129, 130, 144, 279, 303
pump 25, 27, 36, 68, 72, 73, 74, 81, 84, 92, 100, 104, 110, 111, 115, 121, 125, 128, 177, 186, 190, 191, 194, 195, 218, 229, 281, 294, 301
PV panels 118

Q

quantitative easing 78, 265

R
railroads 194
rain barrels 176, 177, 188, 189
reserve currency 12, 270, 274
retirement 3, 9, 18, 51, 60, 77, 117, 253, 256, 257, 259, 272, 286, 295, 296, 297, 304
reverse-osmosis 185
reverse mortgages 262
Robin, Vicki 287, 309, 315
Rocky Mountain Institute 30, 109
root cellar 121, 139, 146, 147, 148, 149, 151, 153, 175, 179, 180, 181, 297
rototiller 159, 160, 170, 197, 266
Royal Canadian Mint 275, 277
Russia 30, 56, 57, 66, 89, 96

S
Saudi Arabia 26, 27, 193, 194
Schwarzenegger, Arnold 42, 53, 55
shotgun 164, 229, 230, 268, 280
shovel 40, 112, 156, 159, 160, 181
silver 10, 77, 105, 228, 262, 263, 266, 270, 271, 273, 275, 276, 277, 278, 279, 280, 281
Simmons, Mathew 26, 27, 29, 84, 193, 194, 195, 306, 314
$mart Power 108, 110, 253
Social Security 20, 87, 264
soft landing 55, 152, 224
Solar Domestic Hot Water 115
Solar Energy International 90
solar panels 7, 61, 63, 72, 73, 89, 93, 108, 109, 118, 119, 120, 121, 122, 125, 126, 127, 128, 130, 131, 132, 145, 200, 201, 204, 244, 245, 247, 261, 266, 276, 281, 289, 292, 293, 302, 303, 304, 305
solar thermal 7, 92, 115, 116, 117, 129, 244
stock market 14, 53, 248, 256, 261, 272, 297
suburbs 64
switchgrass 137

T
tech bubble 16
The Agenda 193, 308
The Association for the Study of Peak Oil & Gas 29
The Blue Zone 217, 296, 309
The Economist 38, 96
The Fulfillment Curve 288

The Gold Standard 10
The Long Emergency 225, 308, 314
The Population Bomb 33
The Progress Paradox 292, 315
The Renewable Energy Handbook 110, 118, 129, 315
Tilford, Monique 288, 309
TIME magazine 55
toilet 72, 73, 85, 153, 172, 183, 186, 187, 188, 190, 269, 292
transit 57, 62, 64, 65, 66, 67, 84, 110, 158, 195, 196
Turner , Garth 77, 307

U
U.S. debt 13
U.S. government 12, 13, 15, 263, 264
unemployment 15, 19, 20, 51, 52, 54, 55, 226, 265, 266
unemployment rate 20, 226
United States Mint 274, 277, 309
utilities 68, 98, 131, 201, 254, 289

V
vegetable garden 2, 71, 93, 146, 156, 161, 176, 266, 286, 294, 305
vegetarian 135, 151, 211, 212, 213, 217
vermi-composter 164

W
Walker, David 14
Wall Street 17, 18, 21, 26, 77, 82, 83, 224, 264, 270, 308
water shortages 41, 42, 43, 44, 45, 49, 176, 188
wind turbines 74, 89, 123, 276
wood heat 101
woodlot 73, 103, 104
woodstove 74, 102, 111, 121, 129, 145, 190, 216, 266
World Policy Institute 37
world population 33, 39
world rice reserves 36

Z
ZENN electric car 200

Recommended Further Reading

PEAK OIL AND ENERGY

Why Your World Is About to Get a Whole Lot Smaller
Jeff Rubin

Twilight in the Desert - *The Coming Saudi Oil Shock and the World Economy*
Matthew R. Simmons

Beyond Oil - *The View from Hubbert's Peak*
Kenneth S. Deffeyes

The End of Oil - *On the Edge of a Perilous New World*
Paul Roberts

The Party's Over: *Oil, War and the Fate of Industrial Societies*
Richard Heinberg

High Noon for Natural Gas:
The New Energy Crisis
Julian Darley

The Hype about Hydrogen: Fact and Fiction in the Race to Save the Climate
Joseph J. Romm

Tar Sands *Dirty Oil and the Future of a Continent*
Andrew Nikforuk

A Thousand Barrels a Second: *The Coming Oil Break Point and the Challenges Facing an Energy Dependent World*
Peter Tertzakian

THE FINANCIAL CRISIS

Empire of Debt: *The Rise of an Epic Financial Crisis*
William Bonner and Addison Wiggin

Mobs, Messiahs, and Markets: *Surviving the Public Spectacle in Finance and Politics*
William Bonner and Lila Rajiva

Bad Money: *Reckless Finance, Failed Politics, and the Global Crisis of American Capitalism*
Kevin Phillips

The Return of Depression Economics and the Crisis of 2008
Paul Krugman

Financial Armageddon: *Protecting Your Future from Four Impending Catastrophes*
Micheal J. Panzner

The Dollar Crisis: *Causes, Consequences, Cures*
Richard Duncan

CHALLENGING TIMES

The Upside of Down: *Catastrophe, Creativity and the Renewal of Civilization*
Thomas Homer-Dixon

The Long Emergency: *Surviving the Converging Catastrophes of the Twenty-First Century*
James Howard Kunstler

Crossing the Rubicon
The Decline of the American Empire at the End of the Age of Oil
Michael C. Ruppert

Emergency: *This Book Will Save Your Life*
Neil Strauss

CLIMATE CHANGE

Carbon Shift: *How the Twin Crises of Oil Depletion and Climate Change Will Define the Future*
Thomas Homer-Dixon

Heat: *How to Stop the Planet From Burning*
George Monbiot

The Weather Makers, *How Man is Changing the Climate and What it Means for Life on Earth*
Tim Flannery

Climate Wars
Gwynne Dyer

PERSONAL FINANCE

Your Money or Your Life
Transforming your Relationship with money and achieving financial independence
Vicki Robin & Joe Dominguez with Monique Tilford

The Progress Paradox:
How Life Gets Better While People Feel Worse
Gregg Easterbrook

RENEWABLE ENERGY AND SUSTAINABILITY

The Renewable Energy Handbook
The Updated Comprehensive Guide to Renewable Energy and Independent Living
William H. Kemp

Biodiesel Basics and Beyond
A Comprehensive Guide to Production and Use for the Home and Farm
William H. Kemp

Ecoholic, *Your Guide to the Most Environmentally Friendly Information, Products, and Services*
Adria Vasil

Recommended Further Reading - continued

Root Cellaring – *Natural Cold Storage of Fruits and Vegetables*
Mike & Nancy Bubel

Saving Seeds – *The Gardener's Guide to Growing and Storing Vegetable and Flower Seeds*
Marc Rogers

Seed to Seed, *Seed Saving and Growing Techniques for Vegetable Gardeners*
Suzanne Ashworth

The All You Can Eat Gardening Handbook *(coming March 2010)*
Cam Mather

WEBSITES

Life After The Oil Crash
www.lifeaftertheoilcrash.net

The Association for the Study of Peak Oil and Gas
www.peakoil.net

The Oil Drum, *Discussions about Energy and our Future*
www.theoildrum.com

Solar Energy International
www.solarenergy.org

The Database of State Incentives for Renewables and Efficiency
www.dsireusa.org

The Office of Energy Efficiency of Natural Resources Canada
www.oee.nrcan.gc.ca.

National Foundation for Credit Counseling
www.nfcc.org

Local Harvest
www.localharvest.org

About the Author

Cam Mather lives independently off-the-electricity-grid with his wife Michelle, using the sun and wind to power his home and business, Aztext Press, which publishes books and DVDs about renewable energy and sustainability. He has produced DVDs on organic vegetable gardening and installing a home-scale wind turbine, and has just written *Thriving During Challenging Times, the Energy, Food and Financial Independence Handbook.* He is currently putting his 30 years of organic and market gardening to use and is writing *The All You Can Eat Gardening Handbook: Easy, Organic Vegetables and More Money in Your Pocket.* This will be released March 2010.

He's working towards the goal of making his home "zero-carbon" and with his extensive garden he aims to be completely food and energy self-sufficient. Through his workshops at colleges he has motivated thousands of participants to invest in energy efficiency and renewable energy to save money, reduce their carbon footprint and to make themselves less dependent on outside sources of energy.

ΛZTEXT PRESS

Environmental Stewartship

In our private lives, the principals in Aztext Press try to minimize our impact on the planet, living off the grid and producing our electricity through renewable energy sources and looking at all facets of our lives, including our personal transportation, heating, cooling, food choices, etc.

We continue to evaluate our books to try minimize their impact, and pick the right mix of recycled stock that still allow photographs to be clear and crisp and offer a quality product to motivate our readers.

In the last 10 years we have planted over 2,000 trees, and each year make a commitment to plant more and nurture existing trees so they thrive and maximize their potential to act like the lungs of the planet and clean the air.

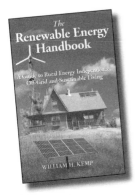

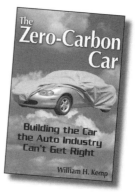

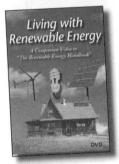